SCHOLAR Study Guide
National 5 Biology

Authored by:
Bryony Clutton (North Berwick High School)
Nikki Haddow (Levenmouth Academy)

Heriot-Watt University
Edinburgh EH14 4AS, United Kingdom.

First published 2021 by Heriot-Watt University.

This edition published in 2021 by Heriot-Watt University SCHOLAR.

Copyright © 2021 SCHOLAR Forum.

Members of the SCHOLAR Forum may reproduce this publication in whole or in part for educational purposes within their establishment providing that no profit accrues at any stage, Any other use of the materials is governed by the general copyright statement that follows.

All rights reserved. No part of this publication may be reproduced, stored in a retrieval system or transmitted in any form or by any means, without written permission from the publisher.

Heriot-Watt University accepts no responsibility or liability whatsoever with regard to the information contained in this study guide.

Distributed by the SCHOLAR Forum.

SCHOLAR Study Guide National 5 Biology

National 5 Biology Course Code: C807 75

ISBN 978-1-911057-85-7

Print Production and Fulfilment in UK by Print Trail www.printtrail.com

Acknowledgements

Thanks are due to the members of Heriot-Watt University's SCHOLAR team who planned and created these materials, and to the many colleagues who reviewed the content.

We would like to acknowledge the assistance of the education authorities, colleges, teachers and students who contributed to the SCHOLAR programme and who evaluated these materials.

Grateful acknowledgement is made for permission to use the following material in the SCHOLAR programme:

The Scottish Qualifications Authority for permission to use Past Papers assessments.

The Scottish Government for financial support.

The content of this Study Guide is aligned to the Scottish Qualifications Authority (SQA) curriculum.

All brand names, product names, logos and related devices are used for identification purposes only and are trademarks, registered trademarks or service marks of their respective holders.

Contents

1 Cell Biology   1

1 Cell structure 3
2 Transport across cell membranes 21
3 DNA and the production of proteins 31
4 Proteins 39
5 Genetic engineering 51
6 Respiration 57
7 Cell biology test 67

2 Multicellular organisms   77

1 Producing new cells 79
2 Control and communication 91
3 Reproduction 109
4 Variation and inheritance 121
5 Transport systems of plants 139
6 Transport systems of animals 153
7 Absorption of materials 173
8 Multicellular organisms test 187

3 Life on Earth   195

1 Ecosystems 197
2 Distribution of organisms 211
3 Photosynthesis 227
4 Energy in ecosystems 239
5 Food production 245
6 Evolution of species 257
7 Life on Earth test 269

4 Appendix 279

A Apparatus and techniques 281
B Laboratory techniques 287
C Field techniques 293

Glossary 298

Hints for activities 304

Answers to questions and activities 305

Cell Biology

1	**Cell structure**		3
	1.1	Cells	4
	1.2	Cell ultrastructure	6
	1.3	Learning points	15
	1.4	Extension materials	16
	1.5	End of topic test	17
2	**Transport across cell membranes**		21
	2.1	Cell membrane structure	22
	2.2	Diffusion	22
	2.3	Osmosis	23
	2.4	Active transport	25
	2.5	Learning points	26
	2.6	Extended response question	27
	2.7	End of topic test	28
3	**DNA and the production of proteins**		31
	3.1	The structure of DNA	32
	3.2	Messenger RNA (mRNA)	34
	3.3	Learning points	35
	3.4	Extended response questions	35
	3.5	Extension materials	36
	3.6	End of topic test	37
4	**Proteins**		39
	4.1	Protein structure and function	40
	4.2	Enzymes	40
	4.3	Factors affecting enzyme activity	43
	4.4	Learning points	46
	4.5	Extended response question	47
	4.6	Extension materials	47
	4.7	End of topic test	49

5 Genetic engineering	51
5.1 Genetic engineering	52
5.2 Learning points	54
5.3 Extended response question	54
5.4 Extension materials	55
5.5 End of topic test	56
6 Respiration	**57**
6.1 ATP	58
6.2 Aerobic respiration	61
6.3 Fermentation	62
6.4 Learning points	64
6.5 Extended response question	65
6.6 End of topic test	65
7 Cell biology test	**67**

Unit 1 Topic 1

Cell structure

Contents

1.1 Cells 4
1.2 Cell ultrastructure 6
1.3 Learning points 15
1.4 Extension materials 16
1.5 End of topic test 17

Learning objective

By the end of this topic you should be able to:

- identify and name the structures found in animal, plant, fungal and bacterial cells;
- describe the functions of the following cell structures:
  - cell wall;
  - mitochondrion;
  - chloroplast;
  - cell membrane;
  - cytoplasm;
  - vacuole;
  - nucleus;
  - ribosome;
  - plasmid;
- state that the cell wall is made of cellulose in plant cells but of different materials in fungal and bacterial cells.

1.1 Cells

Cells are the basic units of life. They come in a variety of shapes and perform many different functions. Cells may be classed as animal, plant, fungal or bacterial.

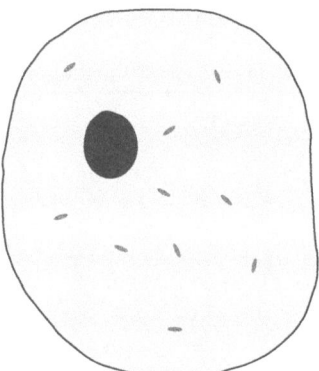

Animal cell

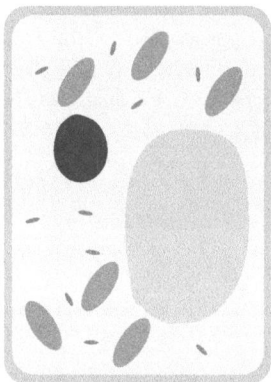

Plant cell

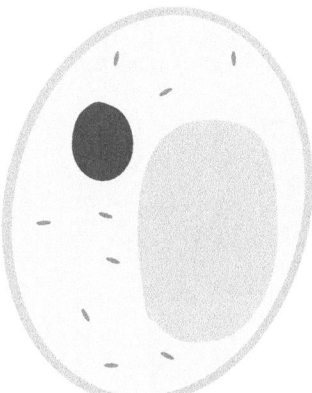
Fungal cell

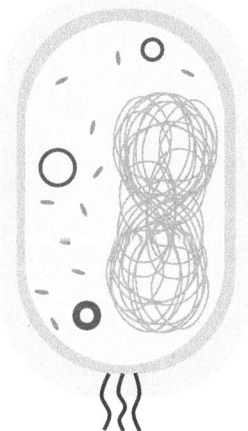
Bacterial cell

Cells contain structures within them called **organelles** which can be seen when cells are viewed under high magnification with powerful microscopes. Each organelle carries out a specific function within the cell. Organelles make up a cell's ultrastructure.

1.2 Cell ultrastructure

The diagram below shows the ultrastructure of an animal cell.

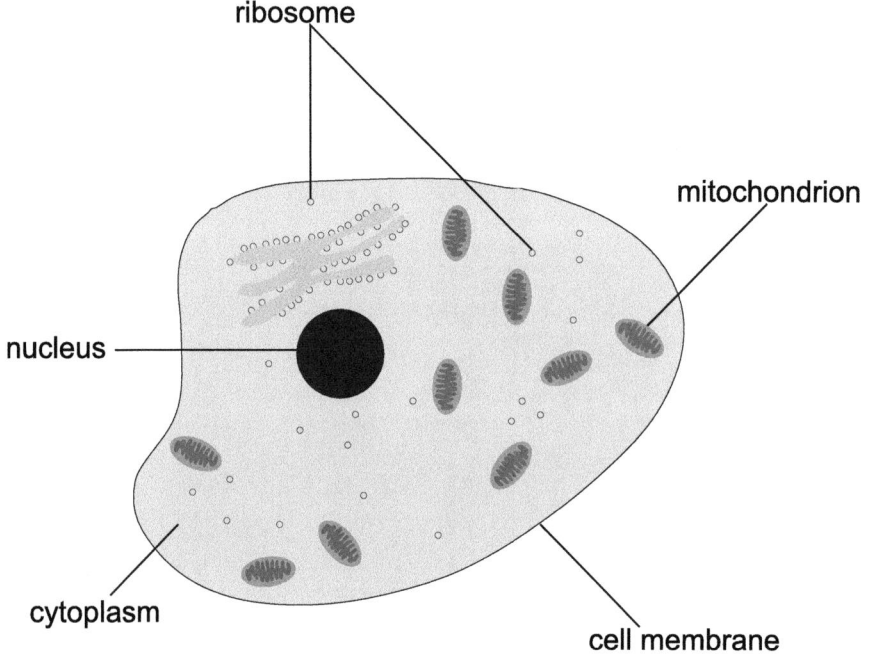

TOPIC 1. CELL STRUCTURE

The diagram below shows the ultrastructure of a plant cell.

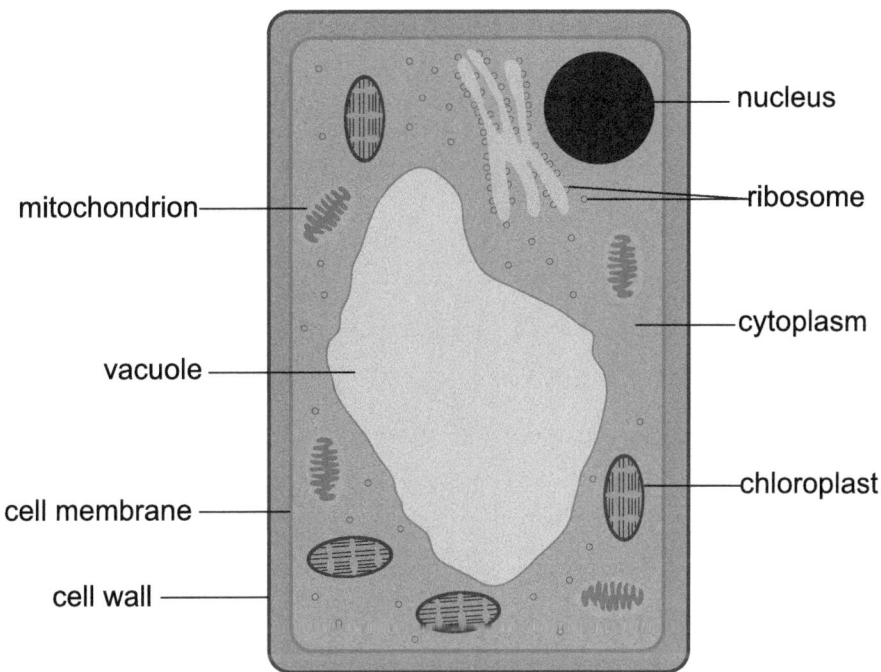

8 UNIT 1. CELL BIOLOGY

The diagram below shows the ultrastructure of a fungal cell.

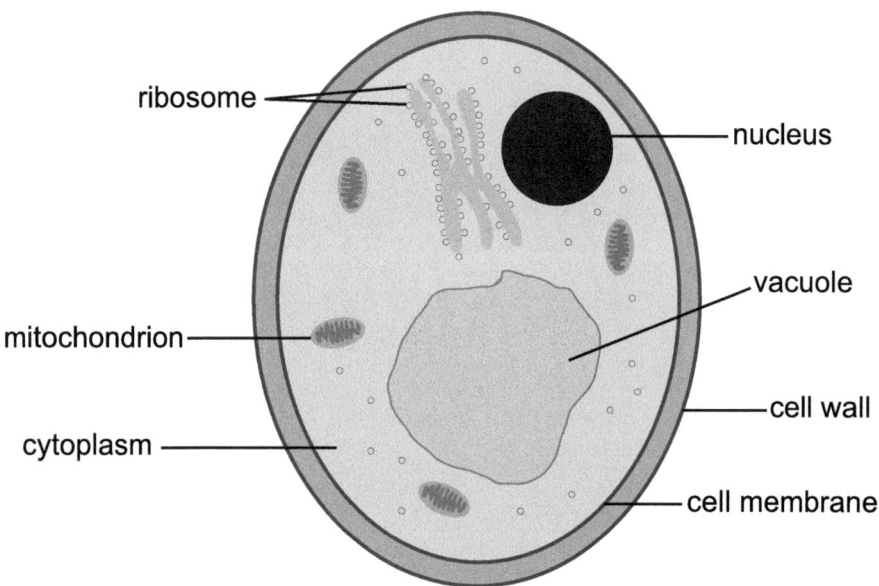

The diagram below shows the ultrastructure of a bacterial cell.

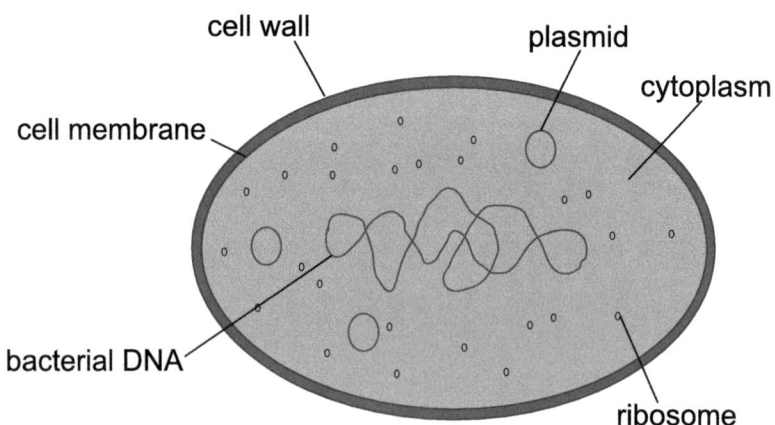

The nucleus is a membrane bound organelle which contains the cell's genetic information (DNA); it controls all activities within the cell. Every cell performs many chemical reactions in order to function, these chemical reactions occur in the cytoplasm. The cell membrane is selectively permeable, it allows small and soluble molecules to enter and exit the cell. Ribosomes are the site of protein synthesis, they may be found free in the cytoplasm or attached to a network of membranes within the cell. Mitochondria (singular mitochondrion) are the site of ATP production by aerobic respiration.

Chloroplasts are the site of sugar production by photosynthesis. A vacuole allows cells to store water, sugar and salts in a solution called cell sap. The cell wall is a layer that surrounds some types of cells, it provides support and gives the cell a rigid structure. The cell wall is made of **cellulose** in plant cells but of different materials in fungal and bacterial cells. A plasmid is a small ring of DNA containing additional genes which are beneficial to the cell for example resistance to antibiotics.

Structure	Function
Nucleus	Controls cell activities
Cell membrane	Controls entry and exit of molecules
Cytoplasm	Site of chemical reactions
Ribosome	Site of protein synthesis
Mitochondrion	Site of aerobic respiration
Chloroplast	Site of photosynthesis
Vacuole	Stores water, sugar and salts in a solution called cell sap
Cell wall	Gives the cell a rigid structure
Plasmid	Contains additional genes which are beneficial to the cell

10 UNIT 1. CELL BIOLOGY

Ultrastructure of an animal cell   Go online

Using the word list, label the following diagram of the ultrastructure of an animal cell.

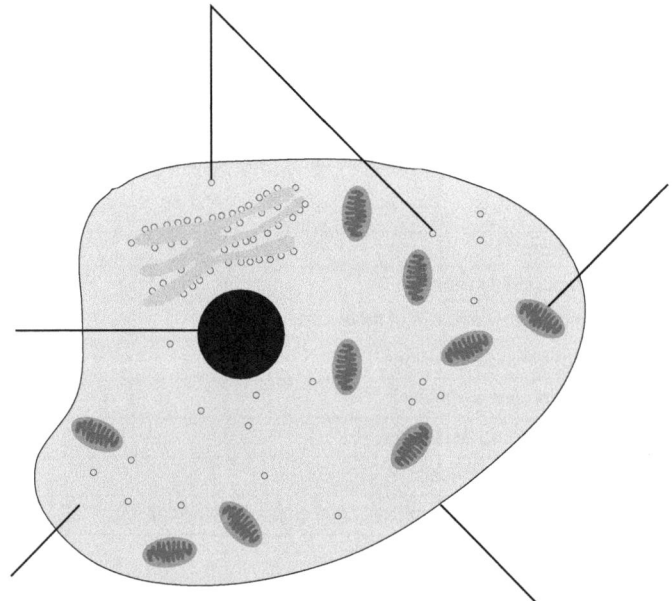

Q1:
Word list: nucleus, cell membrane, cytoplasm, mitochondrion, ribosome, chloroplast, vacuole, cell wall, plasmid, bacterial DNA

TOPIC 1. CELL STRUCTURE

Ultrastructure of a plant cell

Go online

Using the word list, label the following diagram of the ultrastructure of a plant cell.

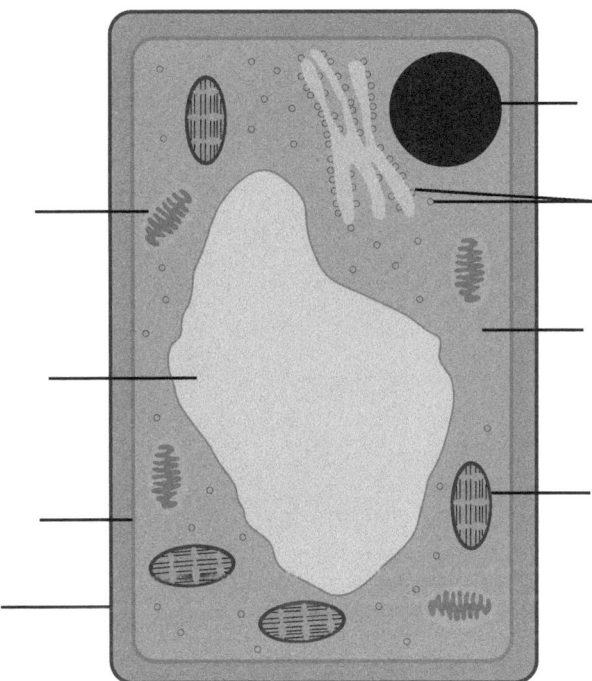

Q2:

Word list: nucleus, cell membrane, cytoplasm, mitochondrion, ribosome, chloroplast, vacuole, cell wall, plasmid, bacterial DNA

12 UNIT 1. CELL BIOLOGY

Ultrastructure of a fungal cell

Go online

Using the word list, label the following diagram of the ultrastructure of a fungal cell.

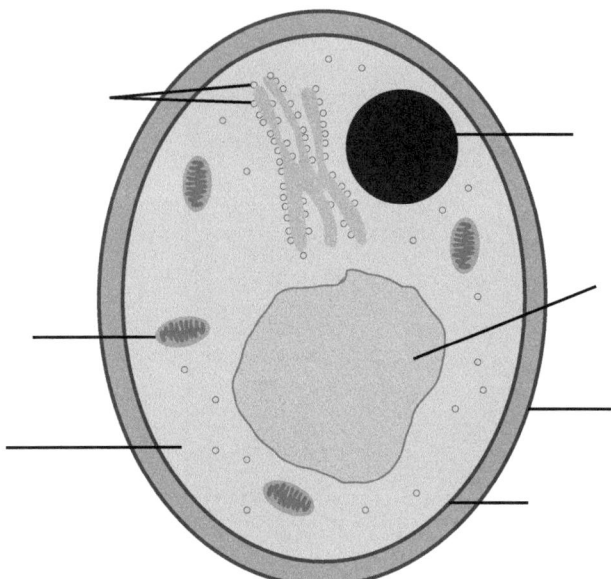

Q3:
Word list: nucleus, cell membrane, cytoplasm, mitochondrion, ribosome, chloroplast, vacuole, cell wall, plasmid, bacterial DNA

TOPIC 1. CELL STRUCTURE

Ultrastructure of a bacterial cell

Using the word list, label the following diagram of the ultrastructure of a bacterial cell.

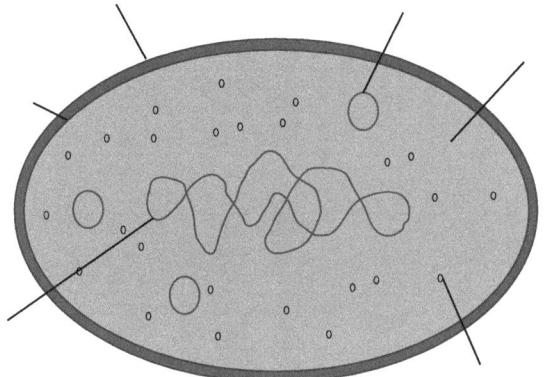

Q4:

Word list: nucleus, cell membrane, cytoplasm, mitochondrion, ribosome, chloroplast, vacuole, cell wall, plasmid, bacterial DNA

Structure and function of a cell

Go online

Match each structure to its function.

Q5:

Structure	Function
Nucleus	Site of chemical reactions
Cell membrane	Site of protein synthesis
Cytoplasm	Controls cell activities
Ribosome	Site of aerobic respiration
Mitochondrion	Gives the cell a rigid structure
Chloroplast	Contains additional genes which are beneficial to the cell
Vacuole	Controls entry and exit of molecules
Cell wall	Stores water, sugar and salts in a solution called cell sap
Plasmid	Site of photosynthesis

Cell structures

Go online

Q6:
Using the word list below, identify the structures found in each type of cell.

Animal cell	Plant cell	Fungal cell	Bacterial cell

Wordlist: nucleus, cell membrane, cytoplasm, ribosomes, mitochondria, chloroplasts, vacuole, cell wall, plasmids

1.3 Learning points

Summary

- Animal cells contain a nucleus, cell membrane, cytoplasm, ribosomes and mitochondria.
- Plant cells contain a nucleus, cell membrane, cytoplasm, ribosomes, mitochondria, chloroplasts, cell wall and vacuole.
- Fungal cells contain a nucleus, cell membrane, cytoplasm, ribosomes, mitochondria, cell wall and vacuole.
- Bacterial cells contain a cell membrane, cytoplasm, ribosomes, cell wall and plasmids.
- The nucleus controls cell activities.
- The cell membrane controls entry and exit of molecules.
- The cytoplasm is the site of chemical reactions.
- Ribosomes are the site of protein synthesis.
- Mitochondria are the site of aerobic respiration.
- Chloroplasts are the site of photosynthesis.
- The vacuole stores water, sugar and salts in a solution called cell sap.
- The cell wall gives the cell a rigid structure.
- The cell wall is made of cellulose in plant cells but of different materials in fungal and bacterial cells.
- Plasmids contain additional genes which are beneficial to the cell.

1.4 Extension materials

An evolutionary theory on the origin of mitochondria and chloroplasts

Lynn Margulis was an American biologist who put forward a theory regarding the origin of mitochondria and chloroplasts known as "endosymbiotic theory". This theory states that mitochondria and chloroplasts were once free-living primitive bacterial cells which were engulfed by a large host cell and eventually (over millions of years) evolved into the organelles we see today. The idea was initially proposed by Russian scientist Konstantin Mereschkowski in 1905 and Lynn Margulis presented microbiological evidence to support the theory in 1967. Several lines of evidence now support the theory:

- mitochondria and chloroplasts are similar in size to bacterial cells
- new mitochondria and chloroplasts are produced by binary fission (splitting in two), a process which bacteria also use to divide.
- the membranes which surround mitochondria and chloroplasts are similar in composition to those of bacterial cell membranes.
- mitochondria and chloroplasts contain circular DNA molecules which are similar to the DNA found in bacteria.

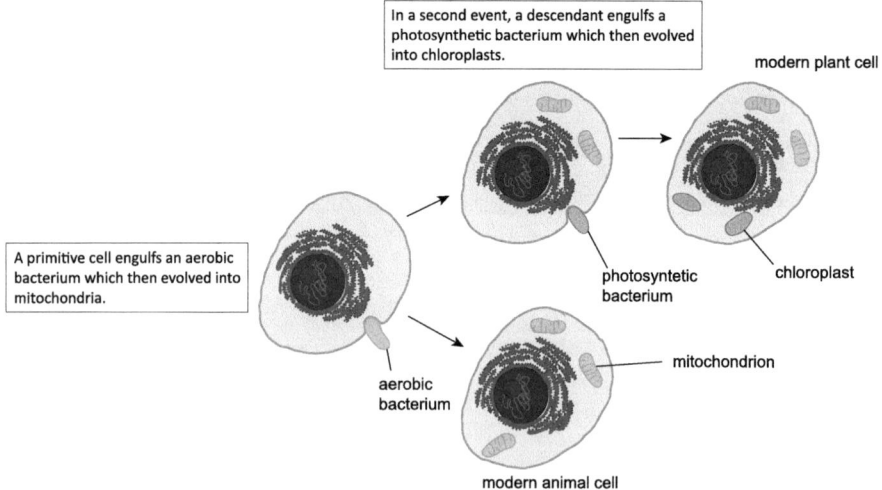

1.5 End of topic test

End of topic test: Cell structure Go online

The list below shows four different cell types:

- Animal
- Plant
- Fungal
- Bacterial

Q7: Identify the cell type(s) which have a nucleus.
....

Q8: Identify the cell type(s) which have chloroplasts
....

Q9: Identify the cell type(s) which have a cell wall.
....

Q10: Describe the function of the nucleus.
....

Q11: Name the structure which controls entry and exit of molecules from a cell.
....

Q12: Name the structure where proteins are synthesised.

18 UNIT 1. CELL BIOLOGY

Q13: Identify each structure shown in the animal cell diagram below.

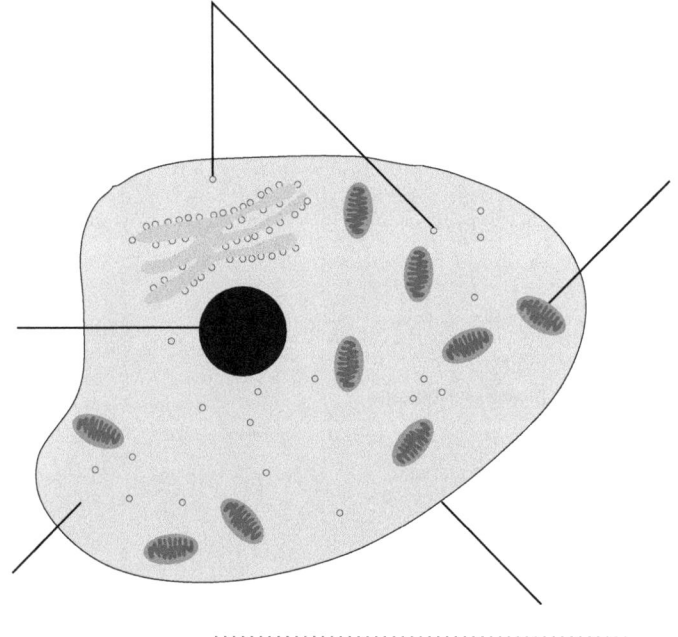

....

TOPIC 1. CELL STRUCTURE

Q14: Identify each structure shown in the plant cell diagram below.

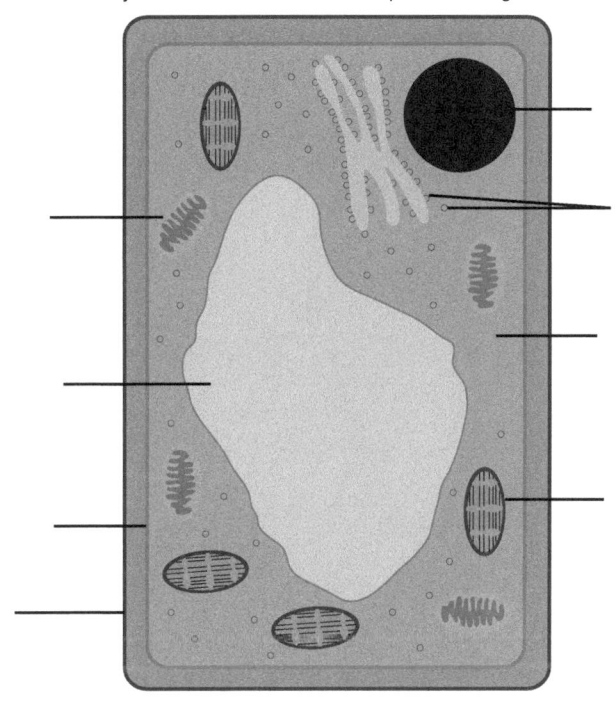

....

Q15: Identify each structure shown in the fungal cell diagram below.

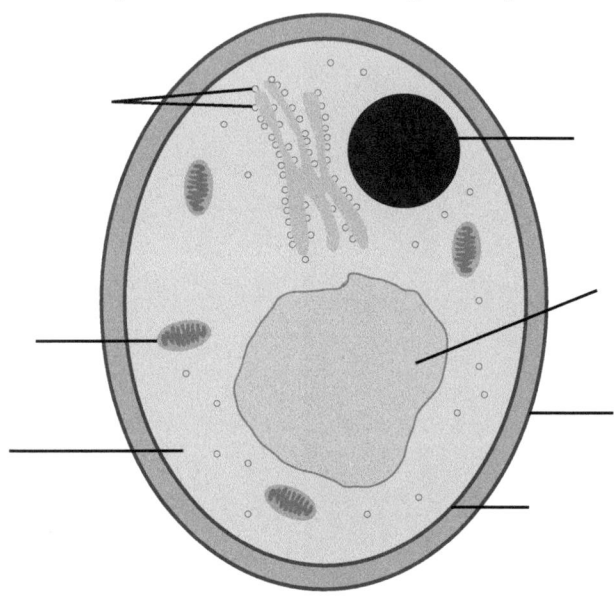

Q16: Identify each structure shown in the bacterial cell diagram below.

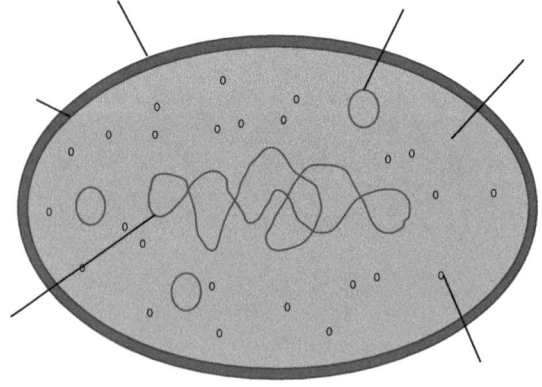

Unit 1 Topic 2

Transport across cell membranes

Contents
2.1 Cell membrane structure 22
2.2 Diffusion 22
2.3 Osmosis 23
2.4 Active transport 25
2.5 Learning points 26
2.6 Extended response question 27
2.7 End of topic test 28

Learning objective

By the end of this topic you should be able to:

- describe the structure of the cell membrane;
- state that the cell membrane is selectively permeable;
- describe two key features of passive transport;
- state that different concentrations of substances exist between cells and their environment;
- give two examples of passive transport;
- describe the process of diffusion;
- describe diffusion of important substances such as glucose, carbon dioxide and oxygen in terms of their concentration gradients;
- describe the process of osmosis;
- describe the effects of osmosis on animal cells when placed in water or a concentrated solution;
- describe the effects of osmosis on plant cells when placed in water or a concentrated solution;
- describe the process of active transport.

2.1 Cell membrane structure

All cells have a membrane which allows molecules to enter and exit the cell. It is selectively permeable, allowing only small and **soluble** molecules to cross. The cell membrane comprises two main components: **phospholipids** and proteins. The phospholipids are arranged in a double layer with proteins interspersed throughout, as shown in the diagram below.

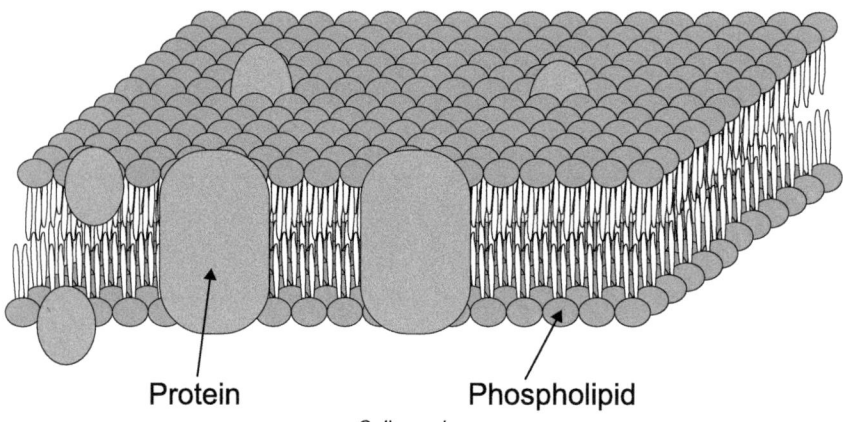

Cell membrane

Some small molecules such as oxygen pass directly through the phospholipid **bilayer** while other larger molecules such glucose are transported across the cell membrane by proteins. Some proteins in the membrane act as receptors and allow the cell to respond to signalling molecules such as hormones.

2.2 Diffusion

Diffusion is the movement of molecules down a concentration gradient from a higher to a lower concentration. A concentration gradient is the difference in concentration of a substance between two areas. A concentration gradient exists between a cell and its external environment as a result of an unequal distribution of molecules across the cell membrane.

Concentration gradient

Diffusion is an example of a passive process as it occurs down a concentration gradient and does not require energy. Diffusion allows cells to gain useful raw materials such as oxygen and glucose;

these useful materials diffuse from a higher concentration outside the cell to a lower concentration inside the cell. Diffusion also allows cells to get rid of waste materials such as carbon dioxide; waste materials diffuse from a higher concentration inside the cell to a lower concentration outside the cell.

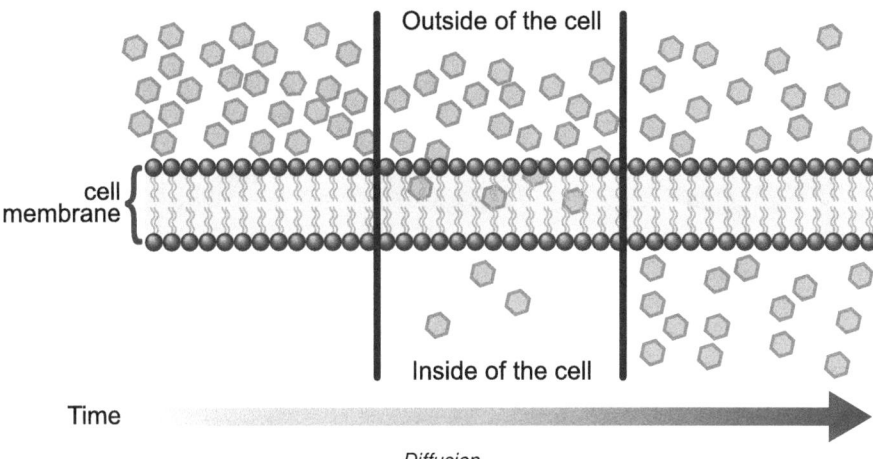

Diffusion

Molecules will continue to diffuse into or out of a cell until the concentration on either side of the cell membrane is equal (the molecules actually continue to move but do so equally in both directions).

2.3 Osmosis

Osmosis is a special type of diffusion involving water molecules. Osmosis is the movement of water molecules from a higher water concentration to a lower water concentration through a selectively permeable membrane. Osmosis is an example of a passive process as it occurs down a concentration gradient and does not require energy.

Osmosis affects animal and plant cells differently. An animal cell placed in a solution with a higher water concentration compared to its cytoplasm (such as pure water) will gain water by osmosis, swell up and burst. An animal cell placed in a solution with a lower water concentration compared to its cytoplasm (such as a strong salt solution) will lose water by osmosis and shrink.

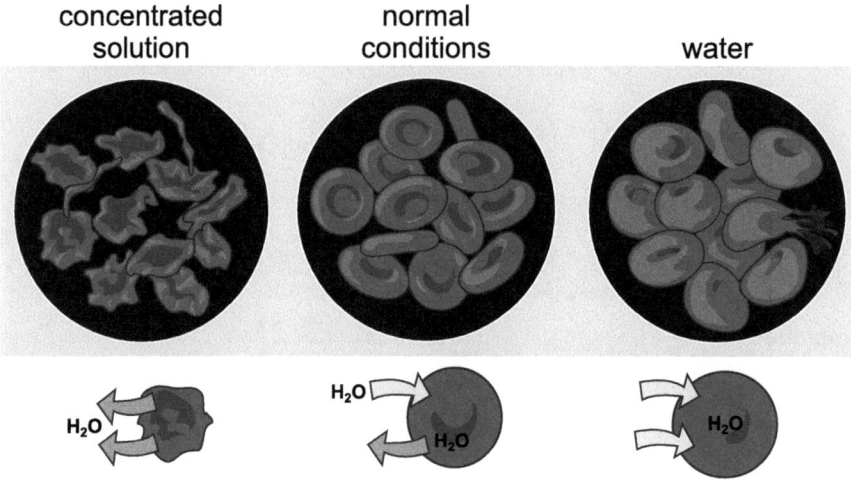

The effects of osmosis on animal cells

A plant cell placed in a solution with a higher water concentration compared to its cytoplasm (such as pure water) will gain water by osmosis and become turgid. A turgid plant cell has a full vacuole and the cytoplasm and cell membrane push up against the cell wall. A plant cell placed in a solution with a lower water concentration compared to its cytoplasm (such as a strong salt solution) will lose water by osmosis and become plasmolysed. A plasmolysed plant cell has a small vacuole and cytoplasm and the cell membrane pulls away from the cell wall.

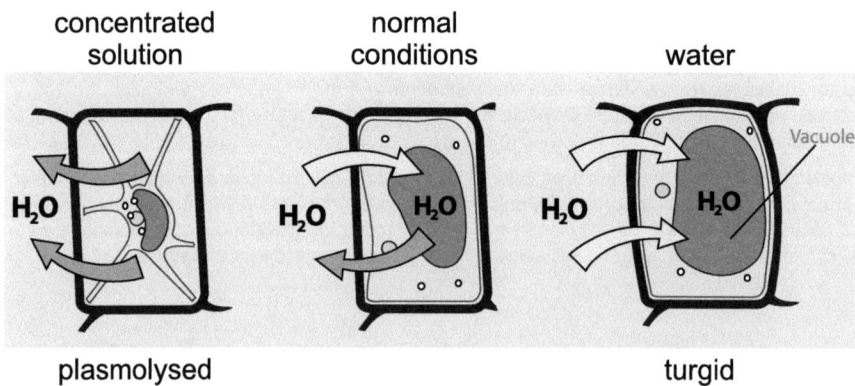

The effects of osmosis on plant cells

TOPIC 2. TRANSPORT ACROSS CELL MEMBRANES

> **Diffusion and osmosis video** Go online
>
> Watch this https://www.youtube.com/watch?v=PRi6uHDKeW4.

2.4 Active transport

Active transport occurs when molecules and ions move from a lower concentration to a higher concentration. This movement occurs against the concentration gradient and therefore requires energy.

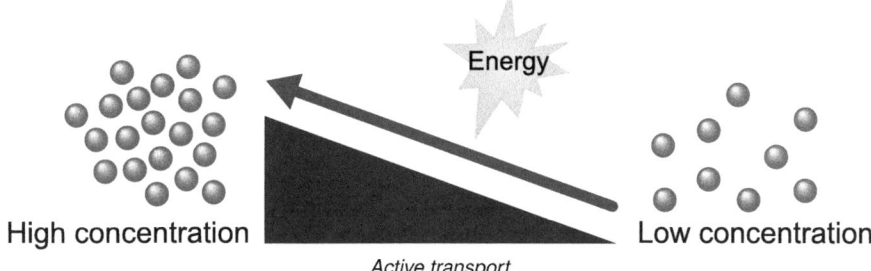

Active transport

Active transport also requires membrane proteins to facilitate the movement of molecules and ions across the cell membrane. One example of active transport is the uptake of ions by root hair cells in plants. Ions are found in a lower concentration in soil water compared to the root hair cells. Plants must therefore use active transport to move these useful ions from a lower concentration in the soil water to a higher concentration in the root hair cells.

© HERIOT-WATT UNIVERSITY

Active transport

Active transport video Go online

Watch this https://www.youtube.com/watch?v=eDeCgTRFCbA.

2.5 Learning points

Summary

- The cell membrane consists of phospholipids and proteins.
- The cell membrane is selectively permeable.
- Passive transport occurs down a concentration gradient and does not require energy.
- Examples of passive transport are diffusion and osmosis.
- Diffusion is the movement of molecules down a concentration gradient from a higher to a lower concentration.
- Glucose and oxygen diffuse into cells from a higher concentration outside the cell to a lower concentration inside the cell.
- Carbon dioxide diffuses out of cells from a higher concentration inside the cell to a lower concentration outside the cell.
- Osmosis is the movement of water molecules from a higher water concentration to a lower water concentration through a selectively permeable membrane.
- An animal cell placed in a solution with a higher water concentration compared to its cytoplasm (such as pure water) will gain water by osmosis, swell up and burst.
- An animal cell placed in a solution with a lower water concentration compared to its cytoplasm (such as a strong salt solution) will lose water by osmosis and shrink.
- A plant cell placed in a solution with a higher water concentration compared to its cytoplasm (such as pure water) will gain water by osmosis and become turgid.
- A plant cell placed in a solution with a lower water concentration compared to its cytoplasm (such as a strong salt solution) will lose water by osmosis and become plasmolysed.
- Active transport requires energy for membrane proteins to move molecules and ions against the concentration gradient.

2.6 Extended response question

Extended response question Go online

A student set up an experiment to investigate the movement of water in to and out of cells. The student used Visking tubing to make a model cell. The experiment was set up as shown in the diagram below and left for 4 hours.

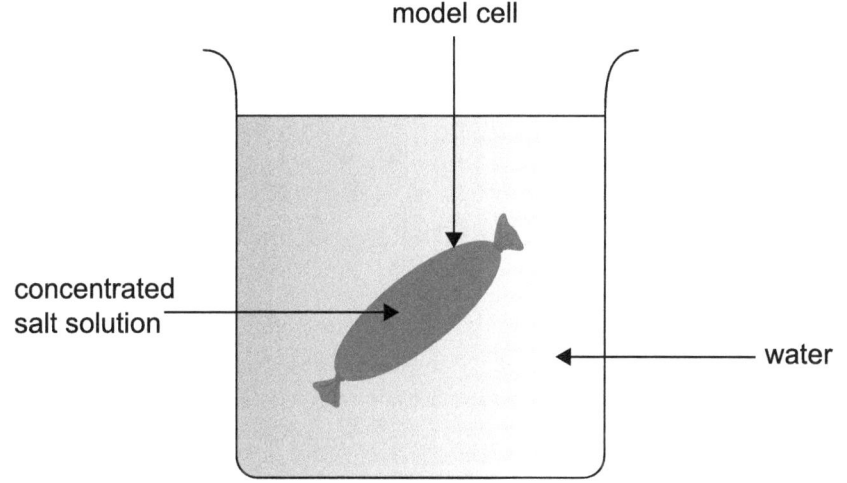

Q1:
Describe the movement of water during the 4 hour time period.

(3 marks)

2.7 End of topic test

End of topic test: Transport across cell membranes Go online

Q2:
The diagram below represents the cell membrane.

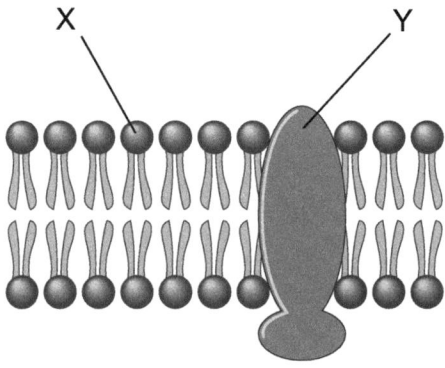

Identify molecules X and Y.

....

Q3: Name the molecules found in the membrane which are required for substances to move in or out of cells by active transport.

The image below shows red blood cells under normal conditions, they were bathed in a solution with a concentration which was the same as the cell contents.

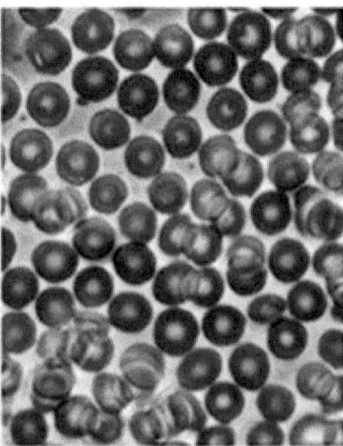

The red blood cells were then immersed in a concentrated salt solution for one hour. The image below shows the effect of the salt solution on the cells.

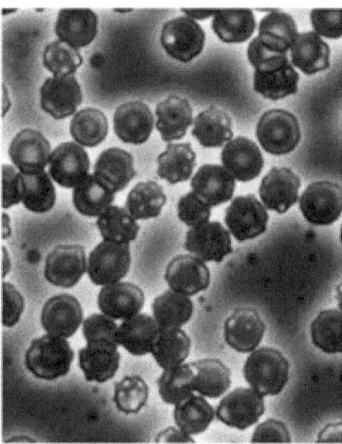

Q4: The change in the red blood cells was caused by the movement of water. Name this movement of water.

....

Q5: Complete the following sentences to describe the process which took place.

Water moved from a (higher/lower) water concentration (outside/inside) the red blood cells to a (higher/lower) water concentration (outside/inside) the red blood cells.

Plant cells from a red onion were also immersed in a concentrated salt solution for one hour. The cytoplasm of these cells contains a purple chemical allowing it to be easily observed. The image below shows a reduced cytoplasm and the membrane of the cell has pulled away from the cell wall.

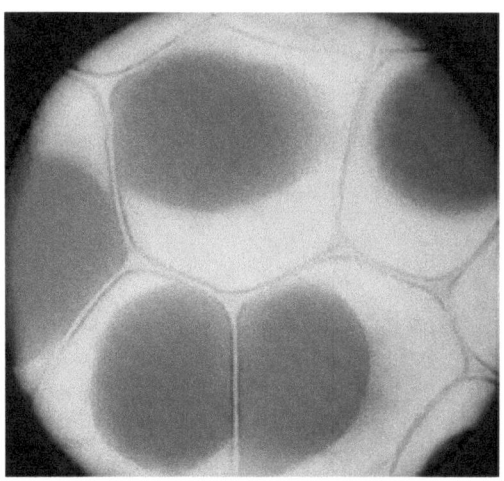

Q6: Name the term which describes the appearance of the cells.

....

Q7: The movement of molecules in or out of cells can be by passive or active transport. List the following descriptions under the correct type of transport.

Passive transport	Active transport

- Requires energy
- Does not require energy
- Moves molecules from a lower concentration to a higher concentration
- Moves molecules from a higher concentration to a lower concentration
- Moves molecules down a concentration gradient
- Moves molecules against a concentration gradient

....

Q8: Give two examples of passive transport.

Q9: Name one substance which diffuses into a muscle cell when it contracts.

....

Q10: Name one substance which diffuses out of a muscle cell when it contracts.

Unit 1 Topic 3

DNA and the production of proteins

Contents

3.1 The structure of DNA 32
3.2 Messenger RNA (mRNA) 34
3.3 Learning points 35
3.4 Extended response questions 35
3.5 Extension materials 36
3.6 End of topic test 37

Learning objective

By the end of this topic you should be able to:

- describe the structure of DNA;
- state that DNA carries the genetic information for making proteins;
- name the four DNA bases;
- describe the base pairing rule;
- state that the base sequence determines amino acid sequence in proteins;
- define the term gene;
- describe the role of messenger RNA (mRNA) in the production of proteins.

3.1 The structure of DNA

DNA is a molecule which carries genetic information. DNA is found in the nucleus of animal, plant and fungal cells coiled up into structures called chromosomes. In bacterial cells the circular chromosome made of DNA is found in the cytoplasm.

The structure of DNA is described as a double-stranded helix. It has two backbones held together by **complementary** base pairs as shown in the diagram below.

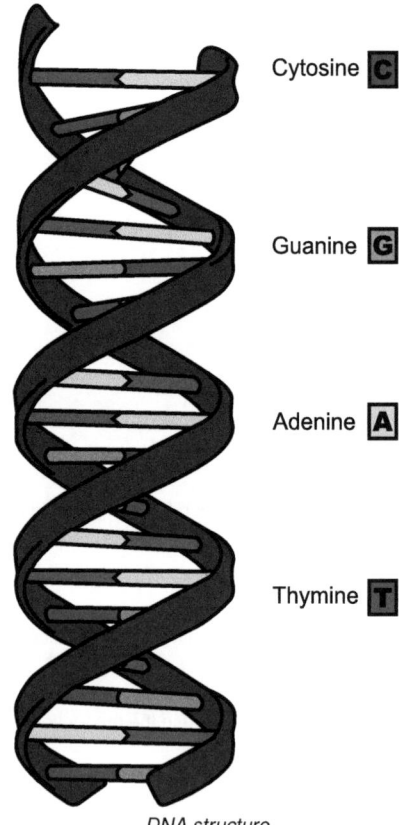

DNA structure

The four DNA bases are adenine, cytosine, guanine and thymine (or A, C, G and T for short). The DNA bases pair up with each other according to the base pairing rule, A is always paired with T and C is always paired with G.

TOPIC 3. DNA AND THE PRODUCTION OF PROTEINS

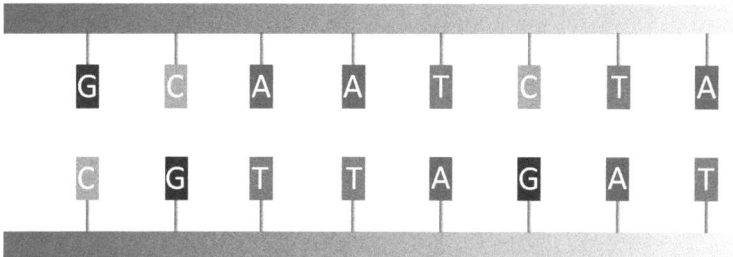

DNA base pairing rule

The four bases make up the genetic code which carries the information for making proteins. A gene is a section of DNA which codes for a protein. Proteins are large molecules which are made up of amino acids. The order of the bases within a gene determines the order of the amino acids in a protein. Proteins allow cells to perform specific functions for example enzymes allow cells to carry out chemical reactions.

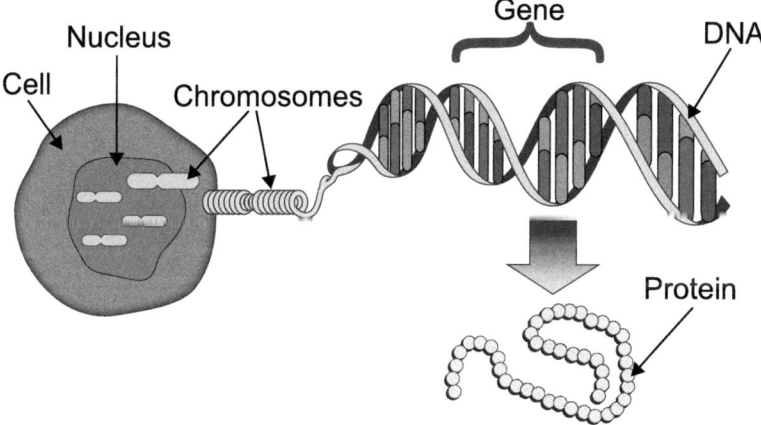

DNA

DNA is a double stranded molecule. The following diagram shows part of one strand. Match the bases into the correct positions to show the complementary strand.

Q1:

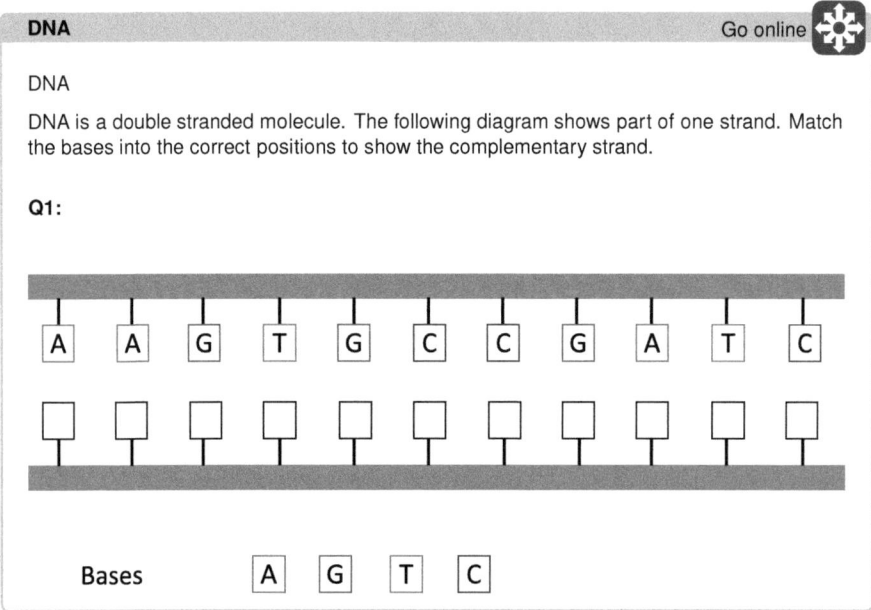

3.2 Messenger RNA (mRNA)

The genetic code containing the information for making proteins is contained within the sequence of bases along the DNA in the nucleus of a cell; however, proteins are assembled at ribosomes in the cytoplasm. A molecule called messenger RNA (mRNA) carries a complementary copy of the genetic code from the DNA, in the nucleus, to a ribosome in the cytoplasm. The order of the bases determines the order of the amino acids in a protein.

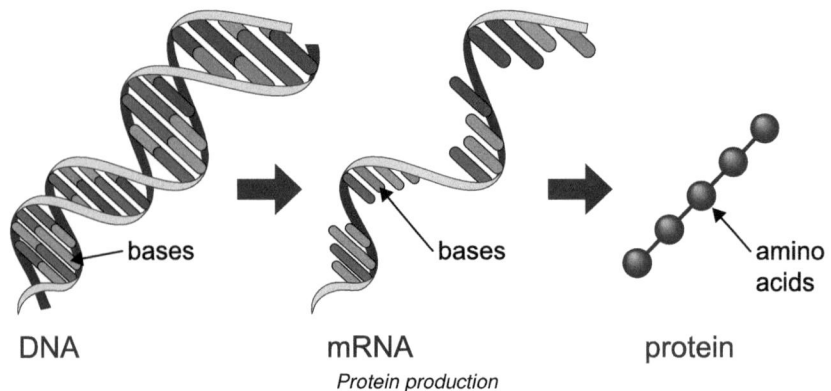

Protein production

3.3 Learning points

Summary

- DNA takes the form of a double-stranded helix held by complementary base pairs.
- DNA carries the genetic information for making proteins.
- DNA contains four bases: adenine, cytosine, guanine and thymine (A, C, G and T).
- DNA bases make up the genetic code.
- The base pairing rule states that adenine (A) is always paired with thymine (T) and cytosine (C) is always paired with guanine (G).
- The base sequence determines amino acid sequence in proteins.
- A gene is a section of DNA which codes for a protein.
- Messenger RNA (mRNA) is a molecule which carries a complementary copy of the genetic code from the DNA, in the nucleus, to a ribosome, where the protein is assembled from amino acids.

3.4 Extended response questions

Extended response questions   Go online

Q2: Describe the structure of DNA.

(3 marks)

....

Q3: Describe the production of protein within a cell.

(4 marks)

3.5 Extension materials

The discovery of the structure of DNA

The structure of DNA was discovered by two scientists, Francis Crick and James Watson, working at the Cavendish Laboratory within the University of Cambridge in 1953. They built molecular models to determine the structure of DNA. Their discovery was based upon evidence provided by other scientists also researching DNA including the knowledge of base pairing provided by Erwin Chargaff. Chargaff discovered that in any DNA molecule the proportion of adenine bases was always the same as the proportion of thymine bases and the proportion of cytosine bases was always the same as the proportion of guanine bases. This rule became known as Chargaff's ratios. Another vital piece of information was an X-ray diffraction image of DNA taken by taken by Rosalind Franklin and Raymond Gosling working alongside Maurice Wilkins. This photo was labelled "photo 51" and suggested that DNA molecules were helical.

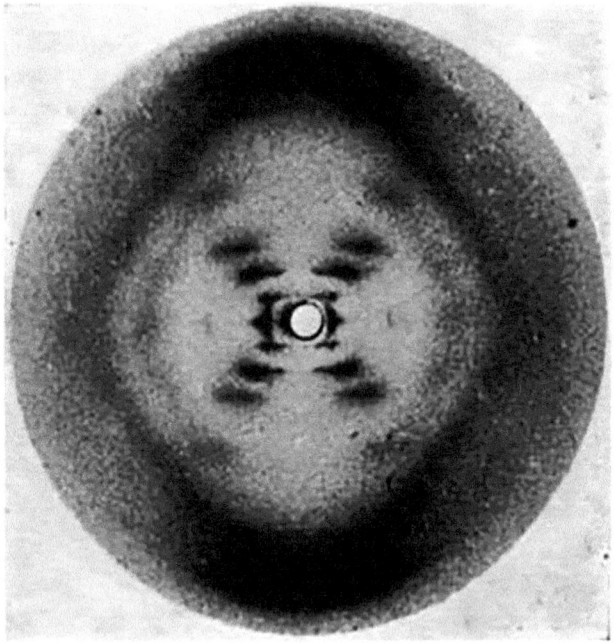

Photo 51

Watson and Crick announced their discovery in a scientific journal called *Nature*. Their article was published alongside others from Maurice Wilkins and Rosalind Franklin which supported their conclusions. Crick, Watson and Wilkins won the Nobel prize in physiology or medicine "for their discoveries concerning the molecular structure of nucleic acids and its significance for information transfer in living material". Franklin was not awarded the prize as she died from cancer four years previously.

3.6 End of topic test

End of topic test: DNA and the production of proteins — Go online

Q4: The following diagram shows part of one strand of DNA. Complete the diagram to show the complementary strand.

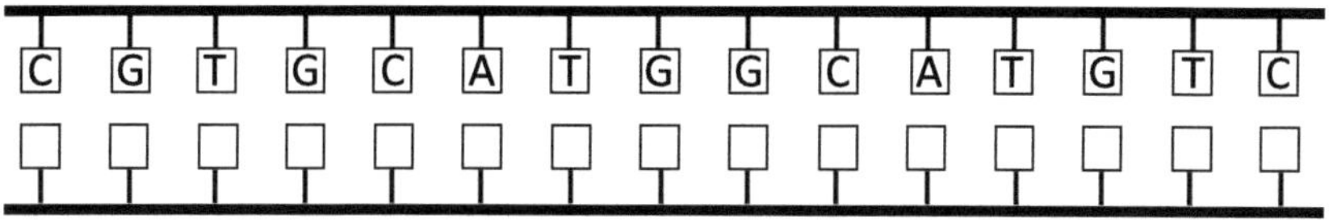

| C | G | T | G | C | A | T | G | G | C | A | T | G | T | C |

....

Q5: Give the full name of each base shown in the diagram above.

- A =
- T =
- C =
- G =

....

Q6: The DNA shown in the diagram above represents a small section of the insulin gene. How would the diagram differ if a section had been taken from a different gene?

....

Q7: DNA carries the genetic information for making _____.

....

Q8: Name the molecule which carries a complementary copy of the genetic code from the DNA in the nucleus to the ribosome in the cytoplasm.

Unit 1 Topic 4

Proteins

Contents

4.1 Protein structure and function 40
4.2 Enzymes 40
4.3 Factors affecting enzyme activity 43
4.4 Learning points 46
4.5 Extended response question 47
4.6 Extension materials 47
4.7 End of topic test 49

Learning objective

By the end of this topic you should be able to:

- state that the variety of protein shapes and functions arises from the sequence of amino acids;
- describe the functions of structural proteins, enzymes, hormones, antibodies and receptors;
- describe the role of enzymes;
- describe the role of the active site of an enzyme;
- give the general word equation for an enzyme mediated reaction;
- describe the features of degradation and synthesis reactions;
- interpret diagrams which illustrate degradation and synthesis reactions;
- give examples of degradation and synthesis reactions;
- explain the term optimum conditions;
- name two factors which can affect the activity of enzymes and other proteins;
- state that enzymes can be denatured, resulting in a change in their shape which will affect the rate of reaction.

4.1 Protein structure and function

There are many different types of proteins within organisms, each doing a different job. Proteins are made up of a variety of 20 different types of amino acid. It is the order of the amino acids which differs between different proteins.

Proteins fold into three-dimensional shapes depending on the amino acids they contain. For example some amino acids are attracted to each other so they try to stick together, others repel water and therefore group together in the middle of a protein away from the surface. Each protein will fold up differently, depending on the order of the amino acids in the chain. A protein must be folded correctly in order to carry out its function.

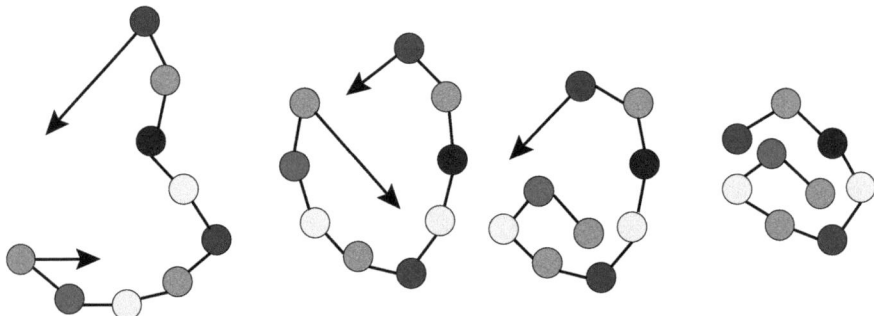

Protein folding

Proteins have many different functions within plants and animals:

- Structural proteins — offer support to the cell/organism.
- Enzymes — speed up cellular reactions and are unchanged in the process.
- Hormones — act as chemical messengers carrying information from one part of the organism to another.
- Antibodies — combine with pathogens to destroy them and protect the body from disease.
- Receptors — are found on the surface of cells and allow signals to be transmitted across the membrane into the cell.

4.2 Enzymes

Enzymes function as biological **catalysts** and are made by all living cells. They speed up cellular reactions and are unchanged in the process. All enzyme reactions follow the same basic word equation:

(molecule the enzyme is working on)   (molecule produced at the end of the reaction)

Enzymes act on their substrate by attaching to them. The area on the enzyme where a substrate attaches is called the active site.

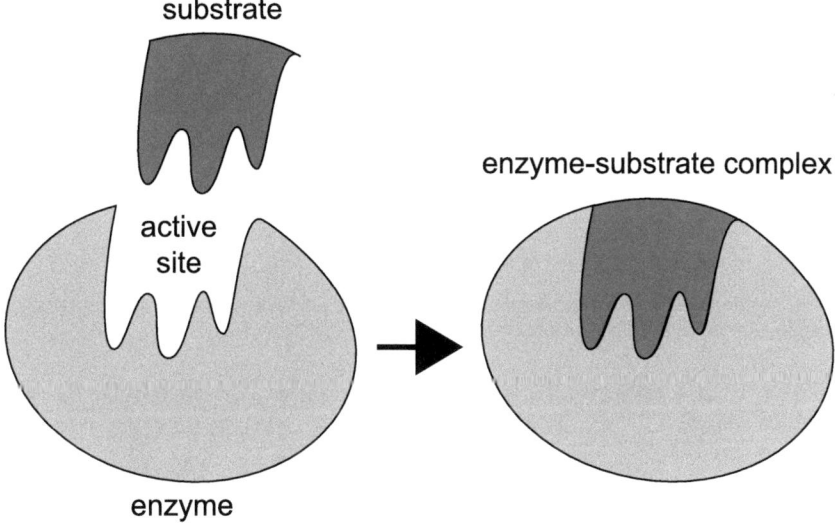

Active site

Each enzyme has a different shaped active site and will therefore bind to different substrate(s). As a result enzymes are specific, they will only catalyse one reaction. The shape of an enzyme's active site is complementary to the shape of its specific substrate(s). This means they can fit together.

Enzymes either catalyse degradation or synthesis reactions. Degradation reactions involve the breakdown of molecules for example the enzyme amylase breaks down starch into maltose.

42 UNIT 1. CELL BIOLOGY

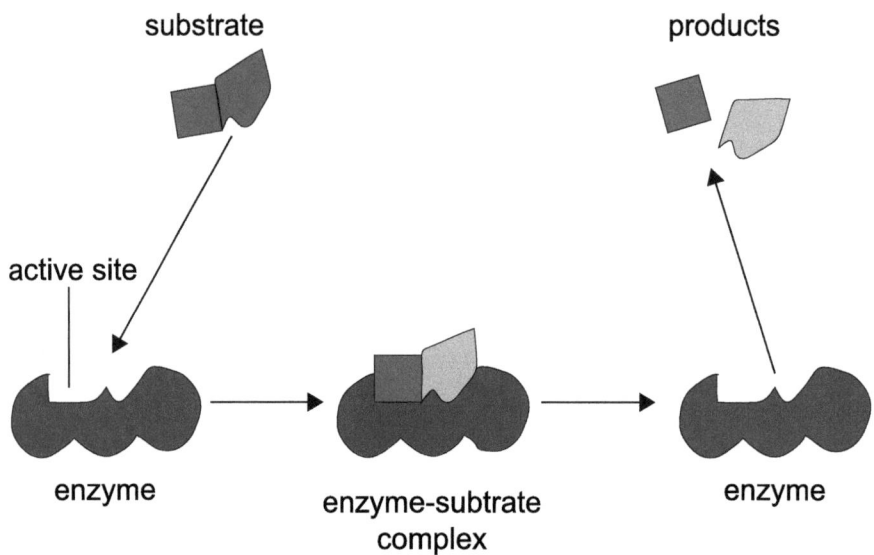

Degradation reaction

Synthesis reactions involve the build-up of molecules for example the enzyme phosphorylase builds up glucose-1-phosphate molecules into starch.

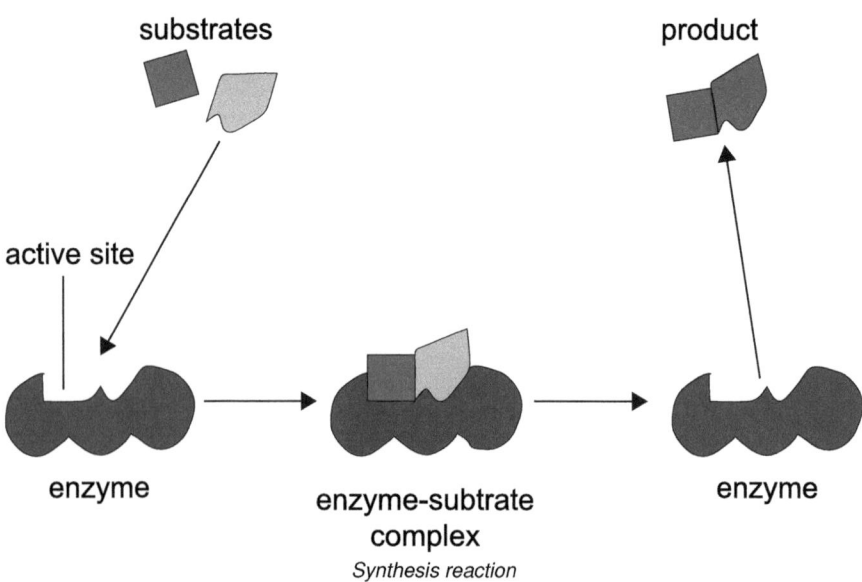

Synthesis reaction

© HERIOT-WATT UNIVERSITY

4.3 Factors affecting enzyme activity

Each enzyme is most active in its optimum conditions. Enzymes are affected by both temperature and pH. As temperature increases enzyme activity also increases until an optimum temperature is reached after which enzyme activity decreases. Human enzymes have an optimum temperature of 37°C.

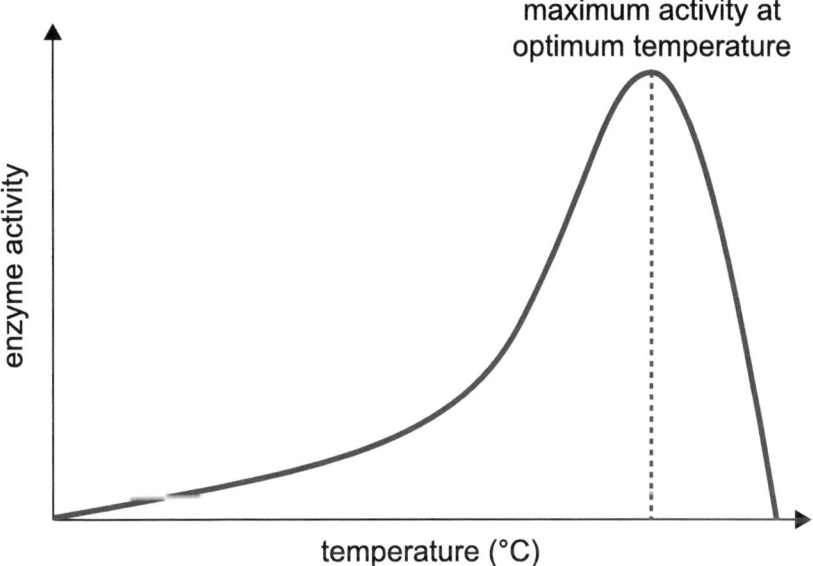

The effect of temperature on enzyme activity

At very high temperatures enzymes become denatured meaning their shape changes so they no longer function properly.

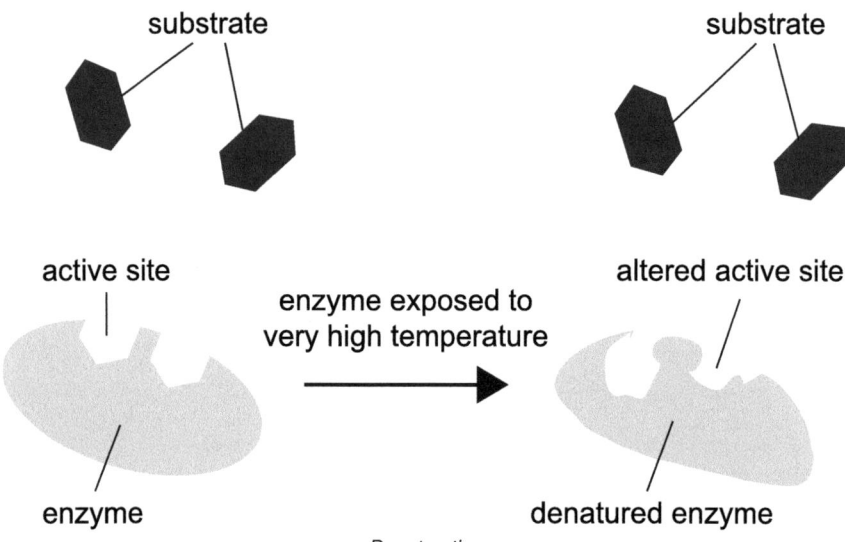

Denaturation

Each enzyme has an optimum pH; the pH where an enzyme is most active. As pH increases enzyme activity also increases until an optimum pH is reached after which enzyme activity decreases. Enzymes within the human body can have drastically different optimum pHs, for example the enzyme pepsin which is found in the stomach has an optimum pH of 2 whereas the enzyme trypsin which is found in the small intestine has an optimum pH of 8.

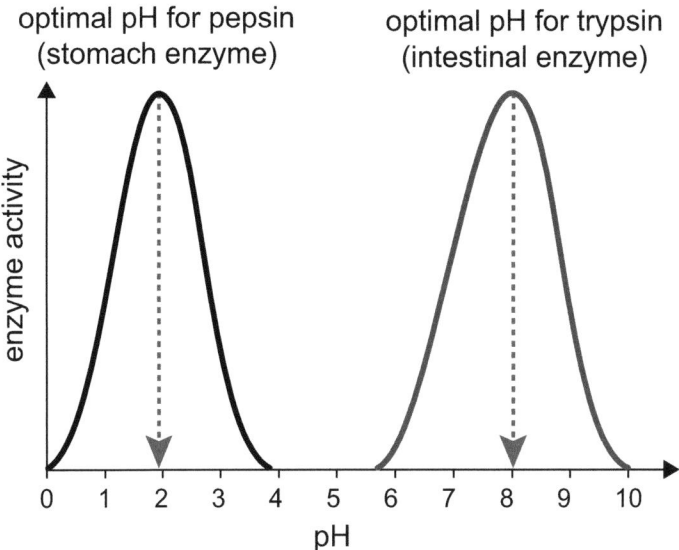

The effect of pH on enzyme activity

4.4 Learning points

Summary

- The variety of protein shapes and functions arises from the sequence of amino acids.
- Structural proteins offer support to the cell/organism.
- Hormones act as chemical messengers carrying information from one part of the organism to another.
- Antibodies combine with pathogens to destroy them and protect the body from disease.
- Receptors are found on the surface of cells and allow signals to be transmitted across the membrane into the cell.
- Enzymes function as biological catalysts and are made by all living cells. They speed up cellular reactions and are unchanged in the process.
- The shape of the active site of an enzyme molecule is complementary to its specific substrate(s).
- Enzyme act upon substrate(s) and enzyme action results in product(s).
- Enzymes can be involved in degradation and synthesis reactions.
- Degradation reactions involve the breakdown of molecules for example the enzyme amylase breaks down starch into maltose.
- Synthesis reactions involve the build-up of molecules for example the enzyme phosphorylase builds up glucose-1-phosphate molecules into starch.
- Each enzyme is most active in its optimum conditions.
- Enzymes and other proteins can be affected by temperature and pH.
- Enzymes can be denatured, resulting in a change in their shape which will affect the rate of reaction.

4.5 Extended response question

Extended response question Go online

The diagram below shows an enzyme and its substrates.

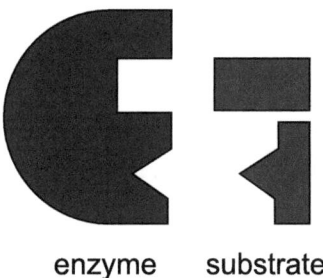

enzyme   substrates

Q1: Using a named example, describe what happens to this enzyme and its substrates during a synthesis reaction.

(4 marks)

4.6 Extension materials

Uses of enzymes in industry

Enzymes can be taken out of organisms, purified and used in industry. One common use in the home is in biological washing powders. Biological washing powders contain amylase, lipases and proteases and break down any stains that contain carbohydrate, fat and protein.

Enzymes are used in the production of many food products. Rennet is an enzyme complex which separates milk into curds and whey. The curds are then used to make cheese. Invertase is an enzyme used to make some soft centred sweets. To make soft centred after dinner mints sucrose, water and the enzyme invertase are mixed to form a thick paste which is covered in chocolate and left to set. The invertase enzyme breaks down the sucrose into a mixture of glucose and fructose which creates a soft centre to the chocolate. Protease enzymes may be used in making baby food. The proteases break down large protein molecules which makes the food easier to digest.

48 UNIT 1. CELL BIOLOGY

Products made using enzymes

4.7 End of topic test

End of topic test: Proteins — Go online

Q2: Match each type of protein to its correct function:

Type of protein	Function
Structural	act as chemical messengers carrying information from one part of the organism to another
Receptor	combine with pathogens to destroy them and protect the body from disease
Enzyme	offer support to the cell/organism
Antibody	speed up cellular reactions and are unchanged in the process
Hormone	allow signals to be transmitted across the membrane into the cell

Biological detergents contain enzymes which help to remove stains from clothes. One type of enzyme found in biological washing powders are proteases. Proteases break down protein molecules into smaller peptide molecules.

Q3: Complete the following sentences by selecting the correct option from each bracket.

The proteases within the biological washing powders are carrying out a (synthesis/degradation) reaction.

In this reaction the proteins are the (substrate/product) and the peptides are the (substrate/product) of the protease enzymes.

....

Q4: Washing machines have different temperature settings. Using your knowledge of enzymes, predict how the cleaning efficiency of a biological washing powder would change if the washing machine temperature setting was changed from 30 °C to 80 °C. Explain your answer:

Prediction: _____ Choose from increase, decrease or stay the same.

Explanation: _____

....

Q5: Other than temperature name one factor which can affect the activity of an enzyme.

50 UNIT 1. CELL BIOLOGY

The diagram below shows three stages in an enzyme catalysed reaction.

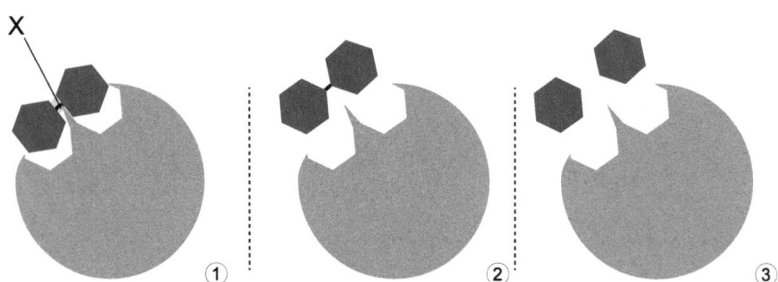

Q6: Complete the following sentences by selecting the correct option from each bracket.

Enzymes (slow down/speed up) chemical reactions, they allow reactions to occur at (lower/higher) temperatures.

Enzymes are (unchanged/changed) by the chemical reaction they catalyse.

....

Q7: Use the numbers in the diagram to complete the boxes below to order the pictures to demonstrate a degradation reaction.

....

Q8: Name site X where the substrate binds to the enzyme.

....

Q9: Some enzymes catalyse synthesis reactions. Describe what is meant by the term "synthesis reaction".

....

Q10: Name the substance of which enzymes are made.

Unit 1 Topic 5

Genetic engineering

Contents

5.1 Genetic engineering 52
5.2 Learning points 54
5.3 Extended response question 54
5.4 Extension materials 55
5.5 End of topic test 56

Learning objective

By the end of this topic you should be able to:

- state that genetic information can be transferred from one cell to another by genetic engineering;
- describe the stages of genetic engineering;
- state that genetic engineering requires enzymes.

5.1 Genetic engineering

Genetic information can be transferred from one cell to another by genetic engineering. This process involves creating a genetically modified organism by transferring genes which result in an improved phenotype. This improved phenotype is often the ability to produce a specific protein. For instance genetic engineering has allowed many **pharmaceutical** products to be produced:

- insulin - used to treat people who have diabetes;
- human growth hormone - used to treat some growth disorders.

The stages of genetic engineering are as follows:

- identify section of DNA that contains required gene from source chromosome;
- extract required gene;
- extract plasmid from bacterial cell;
- insert required gene into bacterial plasmid;
- insert plasmid into host bacterial cell to produce a genetically modified (GM) organism;
- genetically modified bacterial cell divides and produces the required product.

TOPIC 5. GENETIC ENGINEERING

Several stages in this process require enzymes: cutting human DNA to extract the required gene, cutting open the plasmid and sealing the human gene into the bacterial plasmid.

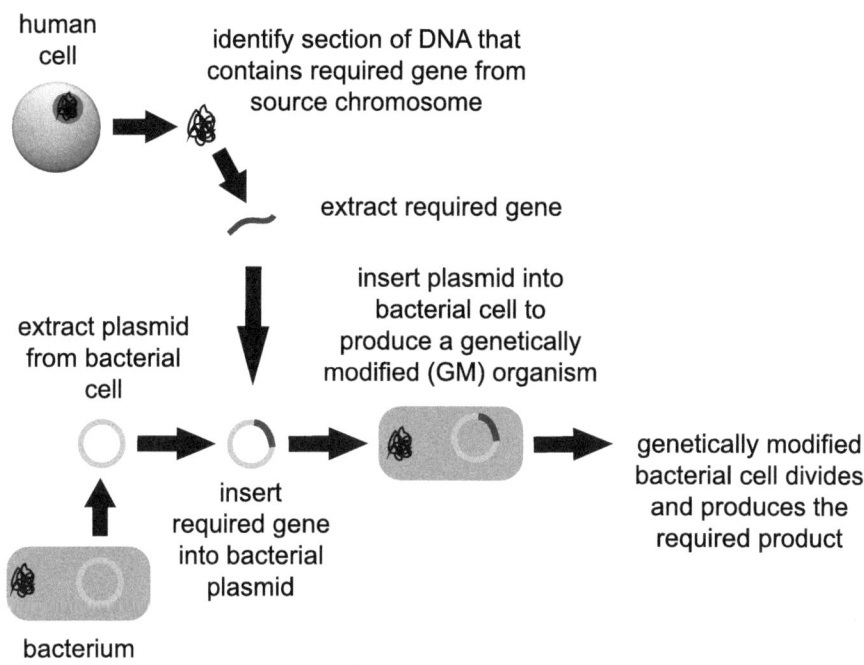

Genetic engineering

5.2 Learning points

Summary

- Genetic information can be transferred from one cell to another by genetic engineering.
- The stages of genetic engineering:
    - identify section of DNA that contains required gene from source chromosome;
    - extract required gene;
    - extract plasmid from bacterial cell;
    - insert required gene into bacterial plasmid;
    - insert plasmid into host bacterial cell to produce a genetically modified (GM) organism.
- The process of genetic engineering requires enzymes to cut human DNA to extract the required gene, to cut open the plasmid and to seal the human gene into the bacterial plasmid.

5.3 Extended response question

Extended response question

Genetic engineering involves the transfer of genetic information from one cell to another. This allows scientists to produce bacterial cells which are capable of making a human protein.

Q1: Describe the stages involved in the genetic engineering process.

(5 marks)

5.4 Extension materials

Applications of genetic engineering

Genetic engineering has applications in many areas of science including medicine, research and agriculture.

Medicine
Many pharmaceutical products are made by genetic engineering, for example insulin. Until the 1980s the only insulin available to diabetics was purified animal-sourced insulin. Although animal-sourced insulin prolonged the life of those with diabetes it caused allergic reactions in some patients. Human insulin was the first pharmaceutical product made by genetic engineering to be mass produced. This insulin is identical to human insulin and conferred a much reduced likelihood of causing allergic reactions. Genetic engineering processes have also allowed scientists to produce other medicines such as human growth hormone, clotting factors and enzymes as well as vaccines and antivenoms.

Research
Genetic engineering is often used by research scientists investigating the role of specific genes within an organism. Loss of function and gain of function experiments involve removing or adding genes from an organism. Scientists then study the effects on the organism and infer the function of the genes of interest.

Agriculture
One of the most controversial areas of science where genetic engineering has been used is the production of genetically modified crops. One such example is golden rice. Golden rice plants have had genes from daffodils and bacteria transferred into their genome which gives it a yellow colour and the ability to produce beta-carotene. When beta-carotene is consumed it is converted into vitamin A which helps to prevent night blindness.

White rice and golden rice (https://en.wikipedia.org/wiki/File:Golden_Rice.jpg by International Rice Research Institute (IRRI), licensed under http://creativecommons.org/licenses/by/2.0/deed.en)

5.5 End of topic test

End of topic test: Genetic engineering   Go online

Q2: The list below describes some of the stages involved in genetic engineering. Fill in the missing stages.

Stage 1: identify section of DNA that contains required gene from source chromosome
Stage 2:
Stage 3: extract plasmid from bacterial cell
Stage 4:
Stage 5: insert plasmid into host bacterial cell to produce a genetically modified (GM) organism

....

Q3: Name the substances required to carry out stages 2 and 4.

The diagram below shows the first stages of genetic engineering.

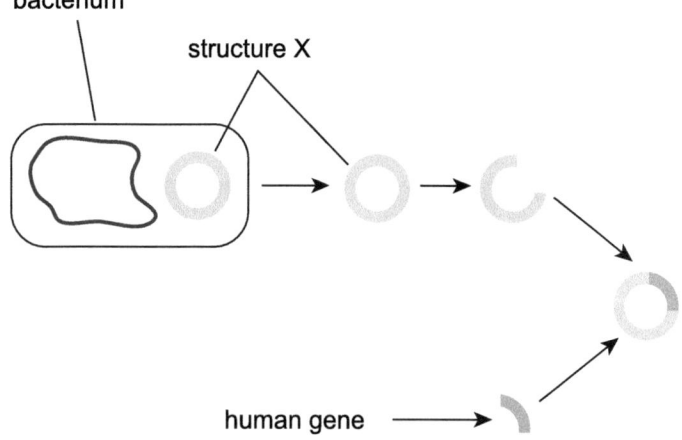

Q4: Name structure X

....

Q5: Describe the next stage in the genetic engineering process.

Unit 1 Topic 6

Respiration

Contents

6.1 ATP 58
6.2 Aerobic respiration 61
6.3 Fermentation 62
6.4 Learning points 64
6.5 Extended response question 65
6.6 End of topic test 65

Learning objective

By the end of this topic you should be able to:

- state that the chemical energy stored in glucose must be released by all cells through a series of enzyme-controlled reactions called respiration;

- state that the energy released from the breakdown of glucose is used to generate ATP;

- describe the role of ATP;

- name four cellular activities which require ATP;

- describe the first stage of respiration;

- describe aerobic respiration;

- describe the fermentation pathway in animal cells;

- describe the fermentation pathway in plant and yeast cells;

- compare the ATP yield of aerobic respiration to that of fermentation;

- state that respiration begins in the cytoplasm;

- state that the process of fermentation is completed in the cytoplasm whereas aerobic respiration is completed in the mitochondria;

- state that the higher the energy requirement of a cell the greater the number of mitochondria present in that cell.

6.1 ATP

The chemical energy stored in glucose must be released by all cells through a series of enzyme-controlled reactions called respiration.

The energy released from the breakdown of glucose is used to generate ATP. ATP stands for adenosine triphosphate; it is a molecule made up of one adenosine group and three phosphate groups. ATP allows energy to be transferred from energy releasing processes (such as respiration) to energy requiring processes (such as cell division).

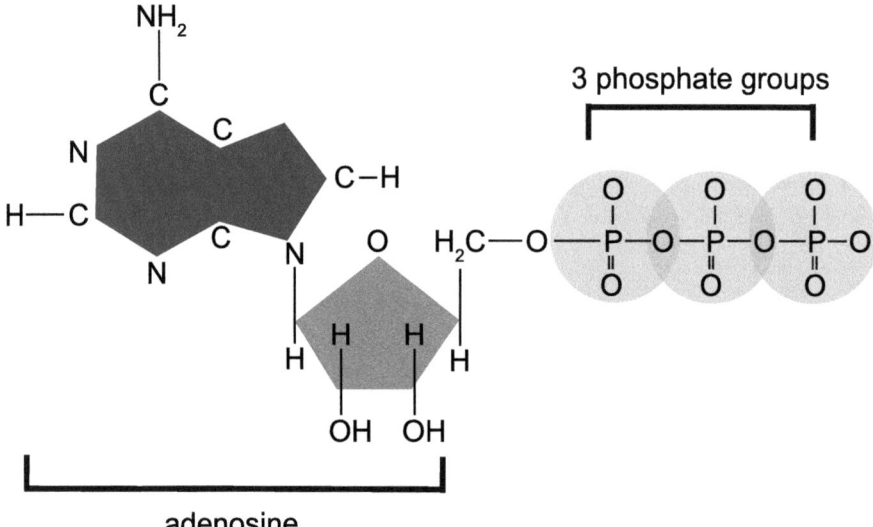

Adenosine triphosphate

The energy transferred by ATP can be used for many cellular activities such as:

- muscle cell contraction;
- cell division;
- protein synthesis;
- transmission of nerve impulses.

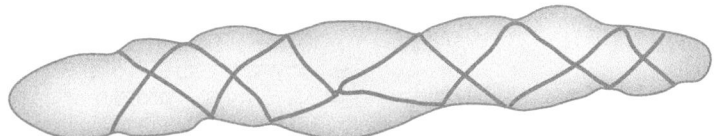

relaxed smooth muscle cell

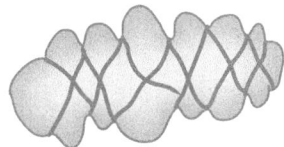

contracted smooth muscle cell
Muscle cell contraction

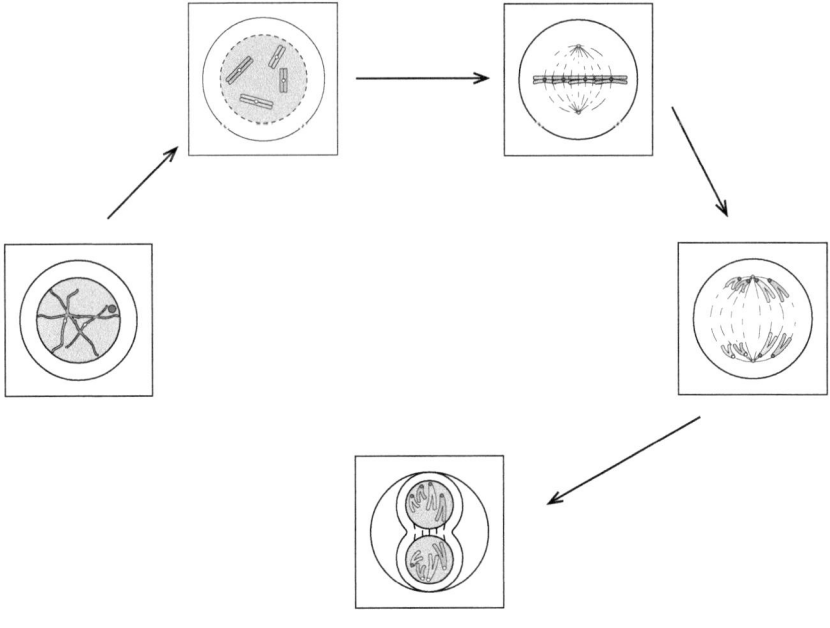

Cell division

60 UNIT 1. CELL BIOLOGY

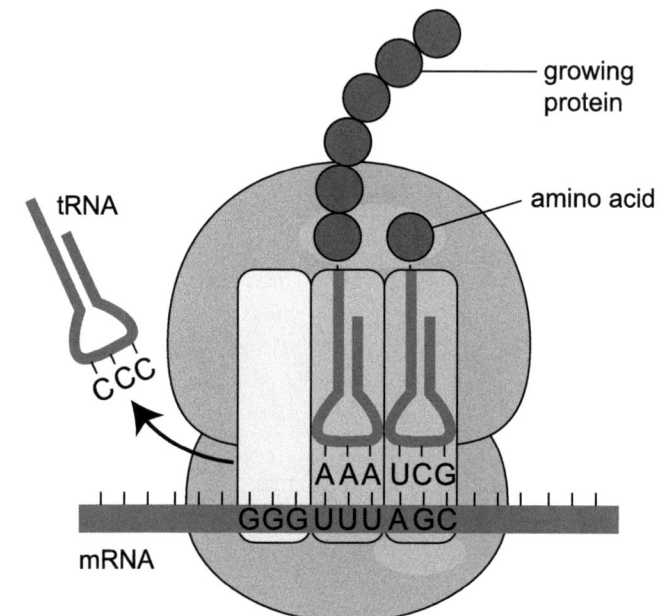
Protein synthesis

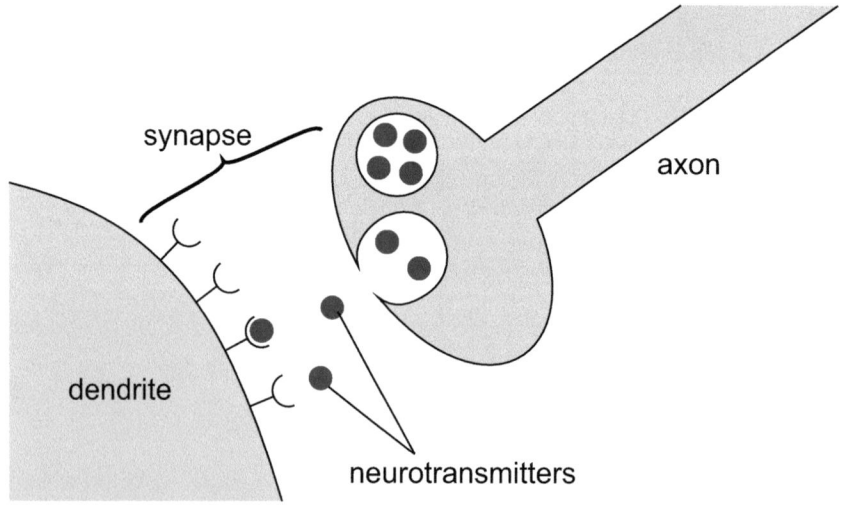
Transmission of nerve impulses

6.2 Aerobic respiration

Aerobic respiration is a chemical reaction which releases the energy stored in glucose, it is described as aerobic as it takes place in the presence of oxygen. The equation below summarises the process of aerobic respiration:

glucose + oxygen → carbon dioxide + water + energy

During aerobic respiration glucose is broken down to two molecules of pyruvate, releasing enough energy to yield two molecules of ATP. This first stage is a series of enzyme controlled reactions which occur in the cytoplasm.

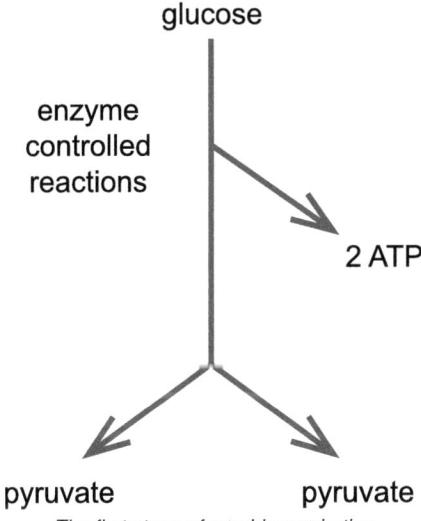

The first stage of aerobic respiration

In the second stage of aerobic respiration, each pyruvate is broken down to carbon dioxide and water, releasing enough energy to yield a large number of ATP molecules. This second stage of aerobic respiration is a series of enzyme controlled reactions which occur in the mitochondria.

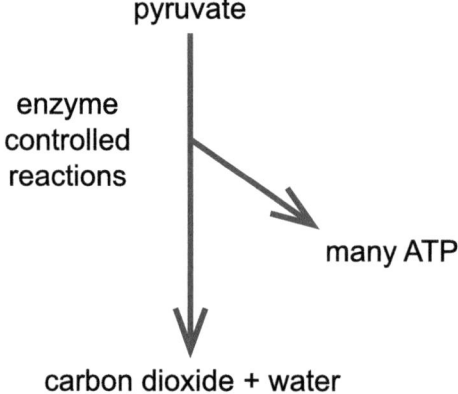

The second stage of aerobic respiration

Since the second stage of aerobic respiration releases the greatest number of ATP and takes place in the mitochondria, the higher the energy requirement of a cell the greater the number of mitochondria present in that cell. For example a muscle cell which requires a lot of energy in order to contract will have a greater number of mitochondria than a skin cell.

6.3 Fermentation

In the absence of oxygen aerobic respiration cannot occur so the fermentation pathway takes place. The first stage of respiration does not require oxygen, therefore the first stage of fermentation is the same as the first stage of aerobic respiration.

During this stage glucose is broken down to two molecules of pyruvate in the cytoplasm, releasing enough energy to yield two molecules of ATP.

During fermentation in animal cells, the pyruvate molecules are converted to lactate in the cytoplasm. The word equation below summarises fermentation in animal cells:

glucose → lactate + energy

During fermentation in plant and yeast cells the pyruvate molecules are converted to carbon dioxide and ethanol in the cytoplasm. The word equation below summarises fermentation in plant and yeast cells:

glucose → carbon dioxide + ethanol + energy

The breakdown of each glucose molecule via the fermentation pathway yields only the initial two molecules of ATP. Therefore aerobic respiration yields many more ATP than fermentation.

Comparison of aerobic respiration and fermentation.

	Aerobic respiration	Fermentation
Oxygen required?	Yes	No
Location of first stage	cytoplasm	cytoplasm
Location of second stage	mitochondria	cytoplasm
Number of ATP produced during first stage	2	2
Number of ATP produced during second stage	Many	None
End products in animal cells	carbon dioxide + water	lactate
End products in plant and yeast cells	carbon dioxide + water	carbon dioxide + ethanol

6.4 Learning points

Summary

- The chemical energy stored in glucose must be released by all cells through a series of enzyme-controlled reactions called respiration.
- The energy released from the breakdown of glucose is used to generate ATP.
- The energy transferred by ATP can be used for cellular activities such as muscle cell contraction, cell division, protein synthesis and transmission of nerve impulses.
- During the first stage of respiration, glucose is broken down to two molecules of pyruvate, releasing enough energy to yield two molecules of ATP.
- If oxygen is present, aerobic respiration takes place, and each pyruvate is broken down to carbon dioxide and water, releasing enough energy to yield a large number of ATP molecules.
- In the absence of oxygen, the fermentation pathway takes place.
- In animal cells, during the fermentation pathway the pyruvate molecules are converted to lactate.
- In plant and yeast cells, during the fermentation pathway the pyruvate molecules are converted to carbon dioxide and ethanol.
- The breakdown of each glucose molecule via the fermentation pathway yields only the initial two molecules of ATP.
- Word summaries of the process of respiration:
    - aerobic respiration: glucose + oxygen → carbon dioxide + water + energy
    - fermentation in plant and yeast cells: glucose → carbon dioxide + ethanol + energy
    - fermentation in animal cells: glucose → lactate + energy
- Respiration begins in the cytoplasm.
- The process of fermentation is completed in the cytoplasm whereas aerobic respiration is completed in the mitochondria.
- The higher the energy requirement of a cell the greater the number of mitochondria present in that cell.

6.5 Extended response question

Extended response question

Q1: Describe the fermentation pathway which takes place in animal cells.

(4 marks)

6.6 End of topic test

End of topic test: Respiration Go online

The diagram below represents the process which takes place in cells to release energy from glucose.

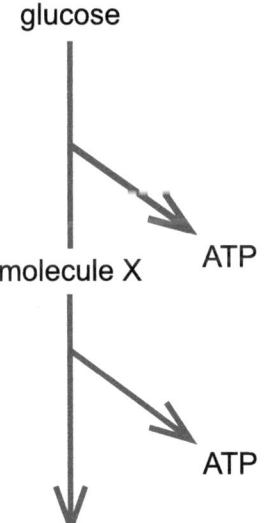

Q2: Name the process shown in the diagram.
....

Q3: Name a cellular process which requires energy.
....

Q4: Name molecule X.

Q5: How many ATP are produced when glucose is converted into molecule X

Q6: Name the substances which control the reactions shown in the diagram above.

Q7: State the location within a cell where molecule X is converted into carbon dioxide and water.

In the absence of oxygen glucose is only partially broken down.

Q8: Name this process.

Q9: Name the product of this process in animal cells.

Q10: State the location within an animal cell where this process takes place.

The following list contains statements relating to aerobic respiration and fermentation.

A. Does not require oxygen
B. Produces pyruvate
C. Yields many ATP
D. Produces carbon dioxide

Q11: Which statements apply to aerobic respiration?

Q12: Which statements apply to fermentation in animal cells?

Q13: Complete the following sentence by filling in the missing organelle:
The higher the energy requirement of a cell the greater the number of _____ present in that cell.

Unit 1 Topic 7
Cell biology test

Cell biology test	Go online

Cell structure

The diagram below represents a cell taken from the root of a plant.

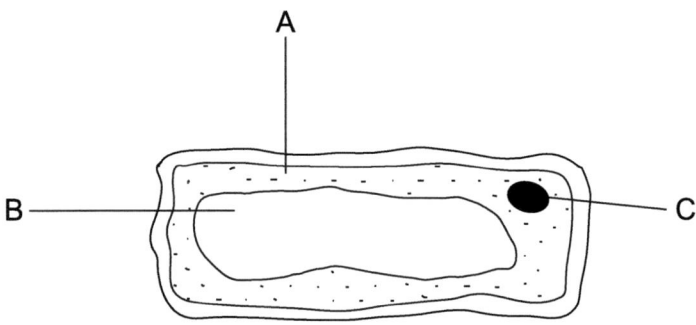

Q1: Name structures A and B.
....

Q2: Give the function of structure C.
....

Q3: Name the structure where protein synthesis takes place.
....

Q4: Describe how the structure of this cell would differ if it were from a green part of a plant leaf.
....

Q5: Name the substance found within plant cell walls.

© HERIOT-WATT UNIVERSITY

Transport across cell membranes

The diagrams below represent plant cells which have been placed in three different solutions.

A 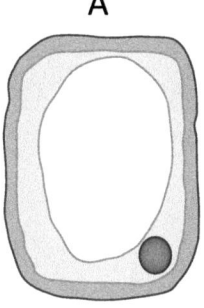 B 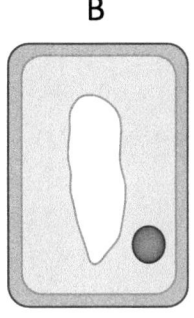 C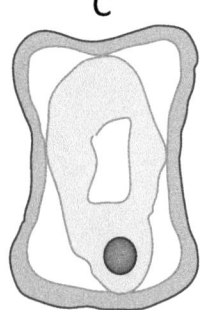

Q6: Which cell was placed in a sugar solution of higher concentration than its cell contents?

....

Q7: What term describes the condition of cell A?

....

Q8: What term describes the condition of cell C?

....

Q9: Explain why an animal cell placed in water will burst.

The diagram below represents a section of the cell membrane.

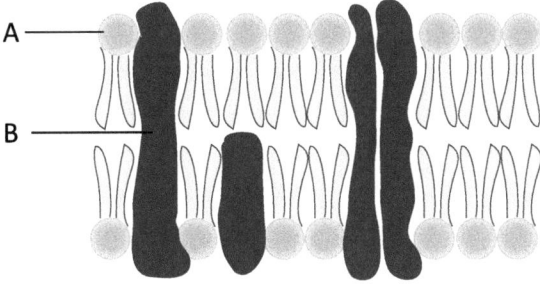

Q10: Name molecules A and B.

....

DNA and the production of proteins

The diagram below represents the production of protein within a cell.

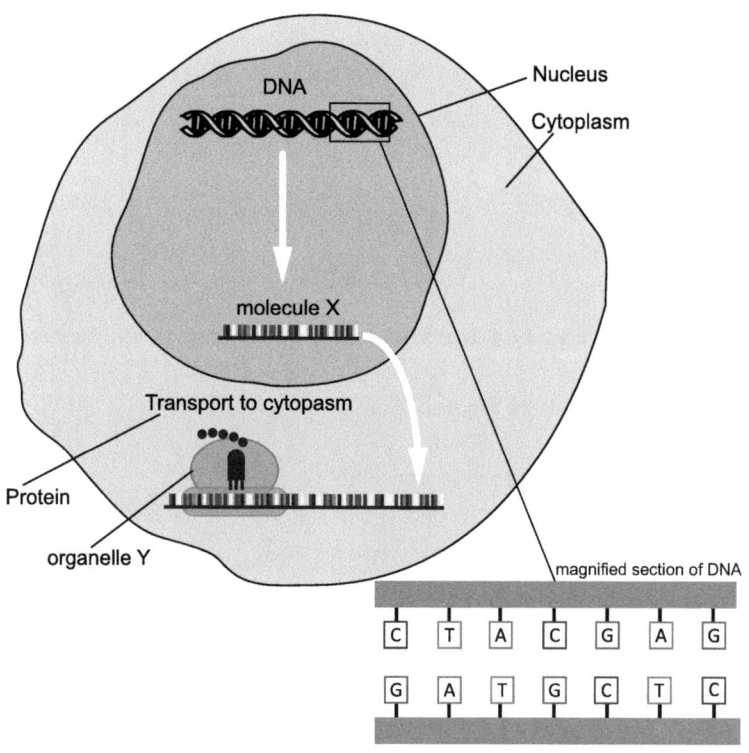

Q11: What name is given to the structure of DNA?

....

Q12: Name molecule X.

....

Q13: Name organelle Y which is the site of protein synthesis.

....

Q14: Give the complementary base sequence missing in the magnified section of the diagram.

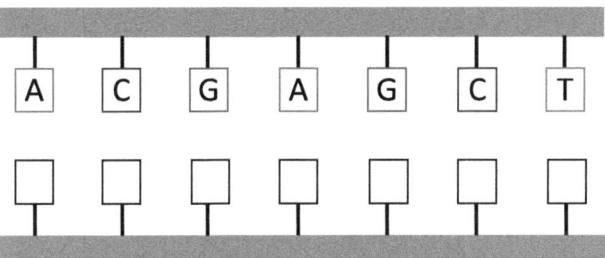

Proteins

Q15: Describe the role of enzymes.

The diagram below shows the action of the enzyme amylase.

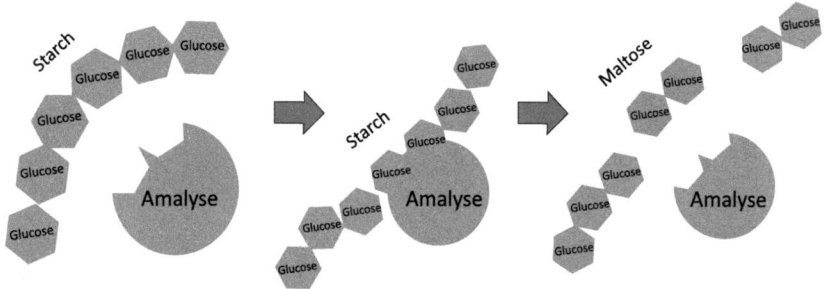

Q16: Name the area on the amylase enzyme where the substrate binds.

....

Q17: What type of reaction is shown in the diagram above?

....

Q18: Explain what would happen to the enzyme if it were heated to 100 °C.

....

Q19: Name one factor which can affect the activity of an enzyme.

....

Q20: What term describes the conditions where an enzyme is most active?

Genetic engineering

The diagram below shows the first stages of genetic engineering.

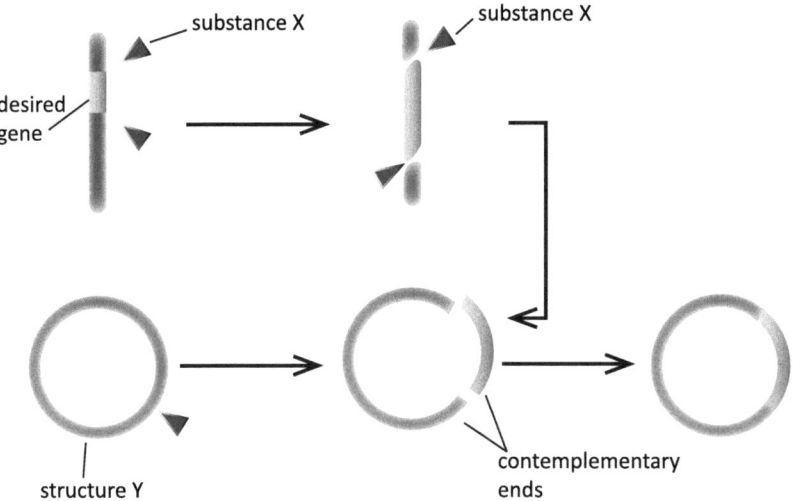

Q21: Name substance X which is used to cut the desired gene out from the human chromosome and cut structure Y open.

....

Q22: Name structure Y

....

Q23: Describe the next stage in the genetic engineering process.

Respiration

The diagram below shows the products of aerobic respiration and fermentation.

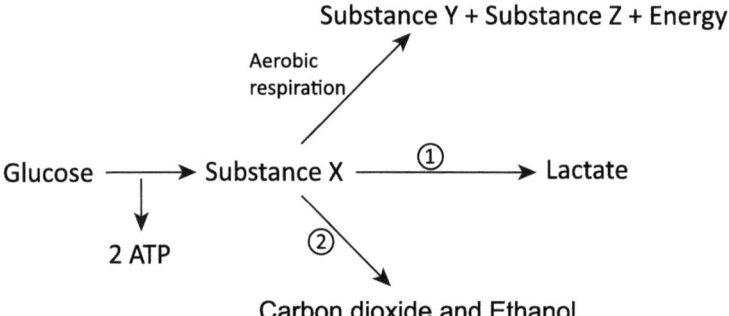

Q24: During the first stage of respiration 2 ATP are produced. Name one cellular process which requires ATP.

....

Q25: State the location within the cell where glucose is converted into substance X.

....

Q26: Name substance X.

....

Q27: Name substances Y and Z.

....

Q28: Which arrow represents fermentation in plant and yeast cells?

Problem solving

A group of students carried out an experiment to investigate the amylase production of different types of bacteria. Amylase is an enzyme which breaks down starch into maltose.

A petri dish was filled with starch agar. Three paper discs were each soaked in a different culture of bacteria and a fourth disc was soaked in water; the discs were placed on the starch agar and the petri dish was incubated at 30°C. After 48 hours the dish was flooded with iodine and the diameter of the clear zone around each disc was measured. The diagram below demonstrates the set-up of the petri dish.

A — B. subtilis
B — E. coli
C — H. hispanica
D — water

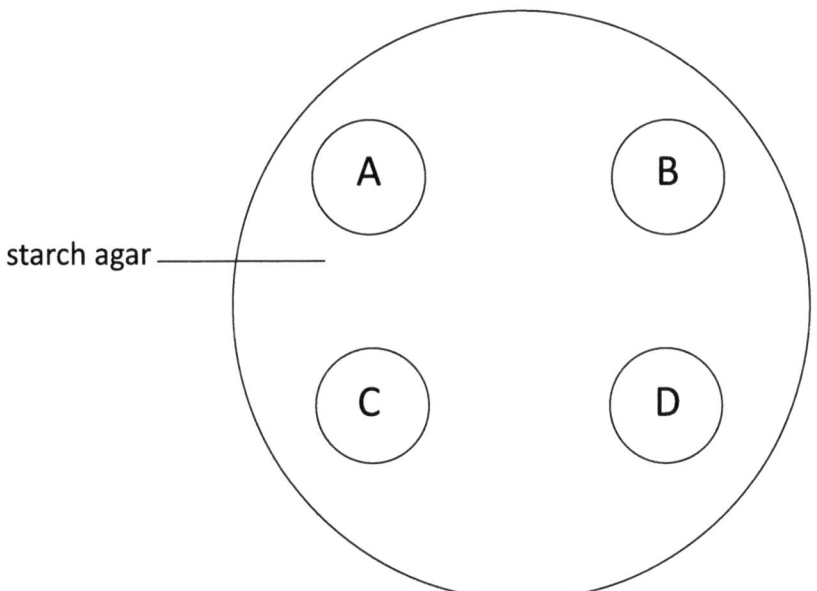

The experiment was repeated twice and the results are shown in the table below.

Disc	Diameter of clear zone (mm)			
	Experiment 1	Experiment 2	Experiment 3	Average
A — B. subtilis	14	9	10	11
B — E. coli	0	0	0	0
C — H. hispanica	9	6	6	
D — water	0	0	0	0

Q29: The disc soaked in water allowed the results from each type of bacteria to be compared with the results when no enzyme is present. Give the term which describes a comparison test like this in a scientific experiment.

....

Q30: Name a variable which must be kept constant to ensure the experiment is valid.

....

Q31: Complete the table by calculating the average diameter of the clear zone for disc C — *H. hispanica*.

....

Q32: State one conclusion from the results of this experiment.

....

Q33: Explain why the experiment was carried out three times and an average calculated.

Multicellular organisms

1	Producing new cells	79
	1.1 Sequence of events of mitosis	80
	1.2 Importance of mitosis	83
	1.3 Stem cells in animals	84
	1.4 Specialisation of cells	86
	1.5 Learning points	88
	1.6 End of topic test	89
2	Control and communication	91
	2.1 Nervous control	93
	2.2 Hormonal control	98
	2.3 Learning points	105
	2.4 End of topic test	106
3	Reproduction	109
	3.1 Chromosomes and gametes	110
	3.2 Reproduction in flowering plants	110
	3.3 Reproduction in animals	114
	3.4 Fertilisation	116
	3.5 Learning points	119
	3.6 End of topic test	119
4	Variation and inheritance	121
	4.1 Variation in species	122
	4.2 Genetic terminology	128
	4.3 Monohybrid crosses	128
	4.4 Phenotypic ratios	134
	4.5 Learning points	135
	4.6 End of topic test	136
5	Transport systems of plants	139
	5.1 Plant organs	140
	5.2 Water transport and the xylem	142
	5.3 The process of transpiration	143

5.4	Sugar transportation and the phloem	145
5.5	Extension work: Comparison of xylem and phloem vessels in plants	146
5.6	Learning points	149
5.7	End of topic test	150

6 Transport systems of animals **153**

6.1	Blood cells	155
6.2	Immune system	157
6.3	Pathway of blood through the body	160
6.4	Structure and function of the heart	162
6.5	Blood vessels	166
6.6	Learning points	169
6.7	End of topic test	169

7 Absorption of materials **173**

7.1	The need for transport	174
7.2	The role of capillary networks	175
7.3	The role of absorption surfaces in the body	176
7.4	Gas exchange in the lungs	177
7.5	Absorption in the small intestine	180
7.6	Learning points	184
7.7	End of topic test	184

8 Multicellular organisms test **187**

Unit 2 Topic 1

Producing new cells

Contents

1.1 Sequence of events of mitosis 80
1.2 Importance of mitosis 83
1.3 Stem cells in animals 84
1.4 Specialisation of cells 86
1.5 Learning points 88
1.6 End of topic test 89

Learning objective

At the end of this topic you should be able to:

- Identify and describe the sequence of events of mitosis.

- Identify chromatids, equator and spindle fibres and state what they are.

- State that mitosis provides new cells for growth and repair of damaged cells and maintains the diploid chromosome complement.

- State that stem cells in animals are unspecialised cells which can divide in order to self renew.

- State that stem cells have the potential to become different types of cell and give examples of these.

- State that stem cells are involved in growth and repair.

- Give examples of the potential uses of stem cells and discuss the ethical issues associated with their use.

- State that specialisation of cells leads to the formation of a variety of cells, tissues and organs.

- State that groups of organs work together form systems and describe the hierarchy that exists.

- Describe a variety of cells from different tissues and relate their structure to their function.

1.1 Sequence of events of mitosis

Chromosomes are thread-like structures found inside the nucleus of every living cell.

Chromosomes carry the genetic information that each species needs to develop, grow and reproduce.

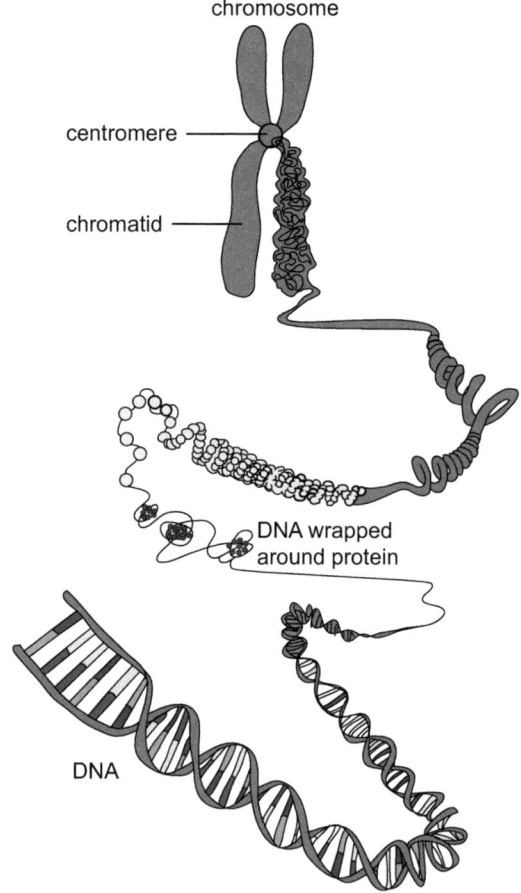

Chromosome structure

At the beginning of cell division, the DNA is replicated, producing two identical copies of DNA. These are connected to each other at the centromere. This replicated X-like structure is now called a sister **chromatid** pair. A 'chromatid' is just one of the strands in the pair.

TOPIC 1. PRODUCING NEW CELLS

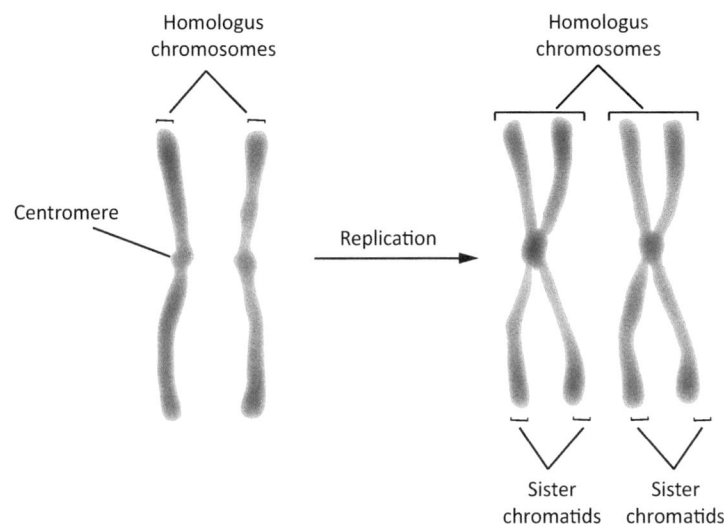

The process of cell division where the nucleus is duplicated is called **mitosis**. Mitosis is required for the growth of an organism and the repair of the organisms' cells. When mitosis takes place, the parent cell duplicates the genetic information contained on its chromosomes before it divides. Each new daughter cell will contain the exact same genetic information as the original parent cell and the organism has an increased number of cells that it can now use to grow in size or to repair damaged cells with.

The process of mitosis takes place in a number of stages.

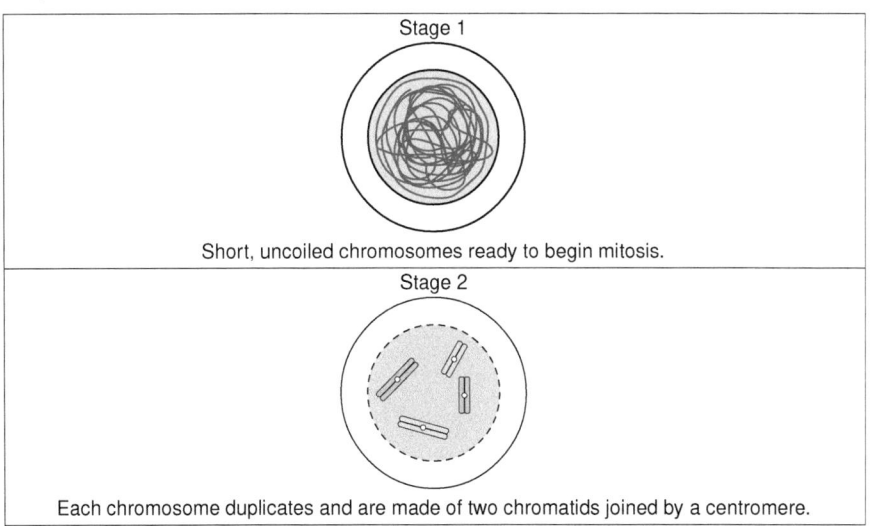

Stage 3
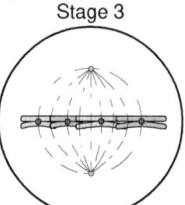
Chromosomes line up along the equator and spindle fibres attach to each pair of chromatids.
Stage 4
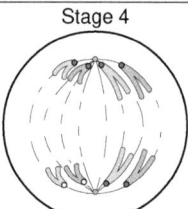
The spindles fibres contract to pull chromatids apart to the opposite poles of the cell.
Stage 5
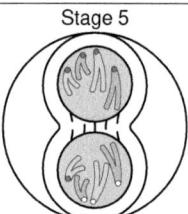
The nuclear membrane will now reform around each group of chromatids and the cytoplasm reforms.
Stage 6
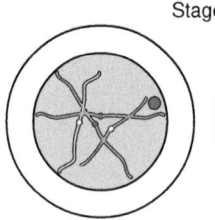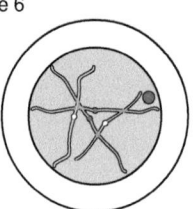
Two new daughter cells are formed. Each have the same number of chromosomes as the original cell.

TOPIC 1. PRODUCING NEW CELLS 83

The sequence of events of mitosis Go online

Q1: Arrange the stages of mitosis in the diagram into the correct sequence.

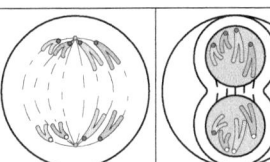 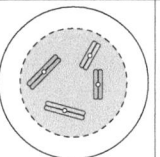 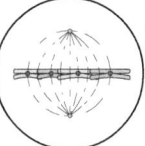 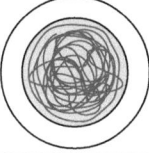

Cell structure Go online

Q2: Which cell structure controls the cells activities?

1.2 Importance of mitosis

Chromosomes are the genetic information found in every living cell. Every species of animal and plant has a characteristic number of chromosomes. This is known as their **chromosome compliment**. When each new cell is formed by mitosis, it is essential that it receives a full complement of chromosomes.

Animal species	Chromosome complement	Plant species	Chromosome complement
Human	46	Cabbage	18
Giant panda	42	Strawberry	14
Dog	78	Carrot	18
Fruit fly	8	Broccoli	18

Importance of mitosis: Questions Go online

Q3: Why is it important that the chromosome compliment in daughter cells is maintained?

....

Q4: Name the term used to describe a cell with a double set of chromosomes.

....

© HERIOT-WATT UNIVERSITY

Q5: The following table shows some information relating to the number of chromosomes in kangaroos. Complete the table, using the information given, to show the chromosome numbers in the remaining cell types.

Kangaroo cell type	Number of chromosomes
sperm	
skin	12
nerve	
zygote	

1.3 Stem cells in animals

Stem cells are unspecialised cells found in animals and they can divide to produce either:

1. more stem cells (self-renew) or
2. cells that develop into **specialised cells**.

Stem cells are involved in growth and repair of body tissues and, depending on the source of the stem cells, have the potential to become various body cells.

Source of stem cells	Where they can be found	Level of potential
Embryonic	Found in embryos	Have the potential to divide and become almost any type of cell in the body
Adult stem cells	Found in **tissues** and **organs** of the body (bone marrow for example)	Have more limitations that embryonic stem cells. They are normally restricted to only being able to form the cell types from the tissue or organ they came from.

Some types of stem cells have been used in medicine for a number of years to repair damaged or diseased organs. Some examples are listed below:

Skin: a rich source of tissue stem cells. Patients with serious burns can be treated using a technique which grows new skin in the lab from skin stem cells.

Blood: a type of stem cell found in the bone marrow is capable of giving rise to all of the different types of blood cells. Bone marrow transplants have been carried out for many years as a treatment for diseases such as leukaemia and other blood disorders.

Wind pipe: scientists have now successfully produced a wind pipe for transplantation using a patient's own stem cells. An artificial windpipe was produced and 'seeded' with the patient's stem cells. After a few days, the wind pipe was ready for use and was transplanted into the patient. This process does not rely on a human donation and will not be rejected by the patient's body.

Stem cells in animals: A study of ethics

The ethical matrix (designed by Professor Ben Mepham, Centre for Applied Bioethics at the University of Nottingham) is a tool to help people analyse an ethical issue and make an informed choice. It is based on three key ethical principles (for further information on these see Mepham: *Bioethics: an introduction for the Biosciences* Oxford University Press (2008), now in 2nd edition):

1. Wellbeing: the safety, welfare and health of an individual or group.
2. Autonomy: an individual's right to be free to choose and make their own decisions.
3. Justice: to what extent a situation is just or fair for an individual or group.

You are asked to consider the following questions with reference to the information in the ethical matrix to help explore your own opinions and feelings using, as far as possible, the evidence here and any other source you feel appropriate.

a) What do you think might be the priority of each of the interest groups?

b) In what way do you think that the three principles apply to each interest group?

c) To what extent do you think others might agree or disagree with you?

d) Might your decision be influenced by the thoughts and beliefs of others?

e) Can you suggest any way round some of the ethical issues that are raised by others?

Interest Groups	Wellbeing (safety, welfare and health)	Autonomy (freedom and choice)	Justice (fairness)
Patients - people who are hoping that stem cell therapies will treat an illness, disease or injury.			
Scientists - people working in stem cell research, developing stem cell therapies to treat patients.			
Embryo - the source of embryonic stem cells for research.			
Society - issues for wider society, such as social priorities, research and medical priorities, and how money should be allocated.			

© HERIOT-WATT UNIVERSITY

Stem cells in animals: Questions

Go online

Q6: What are stem cells?

....

Q7: Name the two main sources of stem cells in animals?

....

Q8: Describe the level of potential of the two named sources of stem cells you identified.

1.4 Specialisation of cells

Multicellular organisms have more than one cell type and are made up of tissues and organs. Organs perform different roles and therefore have cells that are specialised for their function.

The arrangement of this organisational structure is shown in the diagram below.

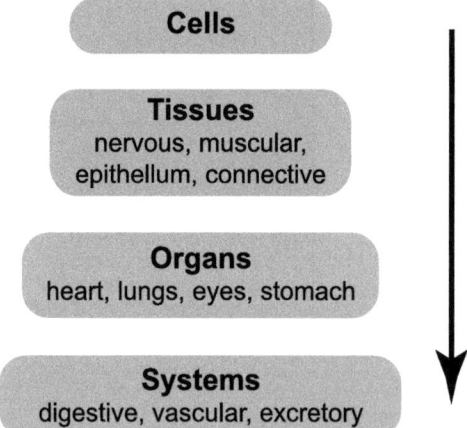

Organisation: cells to tissues to organs to systems

The cells in organs perform different functions and are specialised. This means that they will have a special shape or structure that will help them to carry out their specific role. The following tables give examples of some specialised cells in animals and plants and the role that they play.

Animal cells

TOPIC 1. PRODUCING NEW CELLS

Type of cell	Specialised structure(s)	Function
Red blood cell	Do not have a nucleus and are biconcave in shape to increase the surface area	To transport oxygen around the body
Sperm cell	Have a tail for motility/swimming and lots of mitochondria to provide energy.	To move through the female reproductive **system** and fuse with an egg cell.
Neurone	Have long extensions to send impulses over long distances	To carry information as electrical impulses around the body
Ciliated cell	Hairs called cilia move dirt out of the lungs (respiratory system) or move eggs along the oviduct (reproductive system)	To protect the lungs from dirt and pathogens or transport immotile cells like eggs.

Plant cells

Type of cell	Specialised structure(s)	Function
Phloem cells	The end walls of phloem are like sieves and allow glucose to pass from one cell to the next.	To transport food in plants.
Plant root hair cell	Plant root hair cells have a large projection to increase the surface area that they have for absorbing water.	To absorb water from the soil and aid its entry into a plant.

© HERIOT-WATT UNIVERSITY

88 UNIT 2. MULTICELLULAR ORGANISMS

Specialisation of cells: Questions Go online

Q9: Which of the following shows the terms listed in the correct order of organisation in multicellular organisms?

a) organ → tissue → system
b) tissue → organ → system
c) tissue → system → organ

1.5 Learning points

Summary

- The nucleus of a cell contains all the genetic information for an organism.

- This genetic information is called DNA and is organised into chromosomes.

- The nucleus controls cell activities including the **replication** of chromosomes. This is called mitosis and is the process of DNA copying itself.

- The sequence of events in mitosis begins with chromosomes replicating to become thicker visible pairs of chromatids. The chromosomes then line up along the equator and their chromatids are pulled apart by spindle fibres to opposite poles. Two new nuclei form. After mitosis the cell cytoplasm becomes pinched off and two genetically identical daughter cells are formed.

- Mitosis provides new cells for growth and repair of damaged cells and maintains the diploid chromosome complement.

- Stem cells in animals are unspecialised cells which can divide in order to self renew and are involved in growth and repair.

- Stem cells have the potential to become different types of cell.

- In multicellular organisms specialisation of cells leads to the formation of a variety of cells, tissues and organs. Groups of organs work together form systems.

1.6 End of topic test

End of topic test: Producing new cells Go online

Q10: Stem cells are

a) Specialised cells which are unable to produce new stem cells
b) Specialised cells which have the ability to divide and produce new stem cells
c) Unspecialised which are able to divide and produce new stem cells

....

Q11: The list below describes the stages of cell division. Drag and drop the statements to list them in the correct order.

Pairs of chromatids are pulled apart	Chromosomes move to the equator of the cell
Chromosomes move to the equator of the cell	Nuclear membrane forms
Cytoplasm divides	Chromosomes shorten and thicken
Chromosomes shorten and thicken	Cytoplasm divides
Nuclear membrane forms	Pairs of chromatids are pulled apart

....

Q12:
The diagram below shows a cell undergoing one of the stages of mitosis. Name the part labelled 'A.'

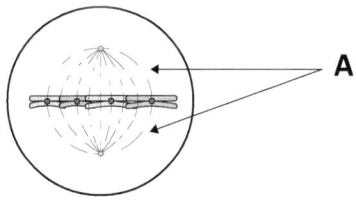

....

Q13: The following diagrams show a cell at four different stages of mitosis. Order them in the right sequence.

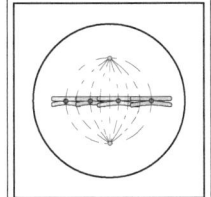 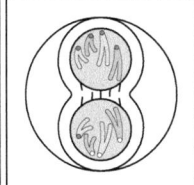 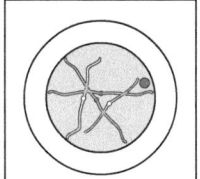 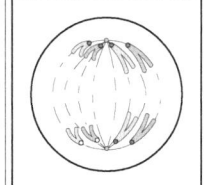

....

Q14: Using the diagram below, describe what is happening at this stage in mitosis.

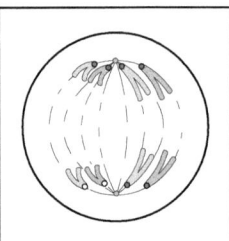

a) Chromosomes line up along the equator and spindle fibres attach to each pair of chromatids.
b) Fibres contract to pull chromatids apart to the opposite poles of the cell.
c) Short, uncoiled chromosomes ready to begin mitosis.
d) Two new daughter cells are formed. Each have the same number of chromosomes as the original cell.

....

Q15: At the beginning of mitosis, there is one original cell. How many times would this original cell have to divide to form 128 cells in total?

Unit 2 Topic 2

Control and communication

Contents

2.1 Nervous control 93
    2.1.1 Micrographs of neurons 93
    2.1.2 Summary animation: Synapse between two neurons 94
    2.1.3 Practical: Reflex actions and the reflex arc 96
    2.1.4 Nervous control: Questions 97
2.2 Hormonal control 98
    2.2.1 Target cell action 100
    2.2.2 Summary animation: Endocrine system animation 100
    2.2.3 Summary animation: Brief overview of parts of the endocrine system and diabetes . . 102
2.3 Learning points 105
2.4 End of topic test 106

UNIT 2. MULTICELLULAR ORGANISMS

Learning objective

At the end of this topic you should be able to:

a) Nervous control

- State that the nervous system consists of central nervous system (CNS) and other nerves.
- State that the CNS consists of brain and spinal cord.
- Identify the structures and state the function of parts of the brain (cerebrum, cerebellum and medulla).
- State that there are three types of neuron (sensory, inter and motor) and describe their role.
- State that receptors detect sensory input/stimuli.
- State that electrical impulses carry messages along neurons and chemicals transfer these messages between neurons, at synapses.
- Describe the structure and function of reflex arc.

b) Hormonal control

- State that endocrine glands release hormones into the bloodstream.
- State that hormones are chemical messengers.
- State that a target tissue has cells with complementary receptor proteins for specific hormones, so only that tissue will be affected by these hormones.
- Describe the roles of insulin, glucagon, glycogen, pancreas and liver in blood glucose regulation.

2.1 Nervous control

The nervous system consists of **central nervous system (CNS)** and other nerves. The CNS consists of **brain** and spinal cord. There are three types of neuron as shown in the table below.

1. **Sensory neurons**: Nerve cells that carry electrical impulses from sense organs to CNS.

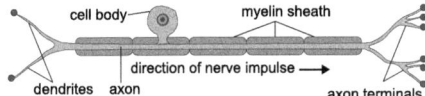

2. **Inter neurons**: Nerve cells that are found in the CNS where they connect with other neurons.

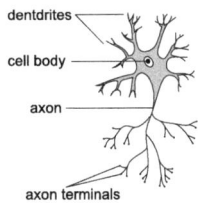

3. **Motor neurons**: Nerve cells that carry electrical impulses from the CNS to muscles and glands (effectors). This can be either a rapid action or a slow response.

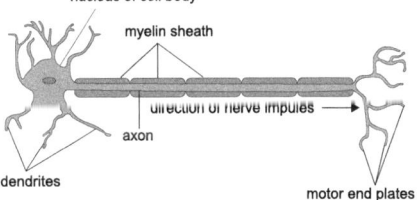

2.1.1 Micrographs of neurons

To examine slides and micrographs of neurons, go to Google images here https://bit.ly/2HUZuz4 .

neurons are specialised cells that carry electrical impulses. neurons have a gap between them, called a **synapse**, that means that they do not actually 'connect' to each other. Neurotransmitters are chemicals which transfer the chemical message of the impulse across this gap between neurons or from a neuron to an effector organ such as an **endocrine gland** or a muscle.

94 UNIT 2. MULTICELLULAR ORGANISMS

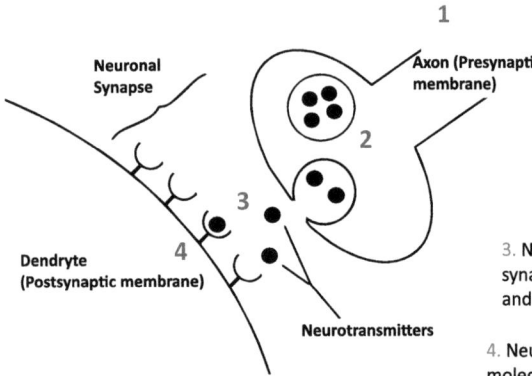

1. Each electrical impulse travels in one direction only (from the cell body to axon terminals).

2. On its arrival, the electrical impulse causes the nerve endings to release the chemical messengers known as neurotransmitters.

3. Neurotransmitters diffuse across the synapse; the gap between the first neuron and the next neuron to it.

4. Neurotransmitters bind with a receptor molecule on the next neuron on the other side of the synapse and this neuron is now stimulated to transmit the electrical impulse.

2.1.2 Summary animation: Synapse between two neurons

| Synapse between two neurons | Go online |

Watch this https://youtu.be/H_81gwAnjDU.

The cerebrum is the section of the brain that controls memory, conscious thoughts, intelligence and emotions. It is the largest part of the brain and is divided into two hemispheres (halves) that receive and process information from the sensory neurons. The cerebrum will then send impulses, via the motor neurons, to bring about the appropriate response to this information.

The **cerebellum** is the section of the brain that controls coordination, movements and balance. The cerebellum is located at the back of the brain just underneath the **cerebrum**.

The **medulla** is the section of the brain that controls breathing and heart rate. The medulla can be found just above the spinal cord.

Interactivity on parts of the brain	Go online

Q1: Use the wordlist to name the parts of the brain:

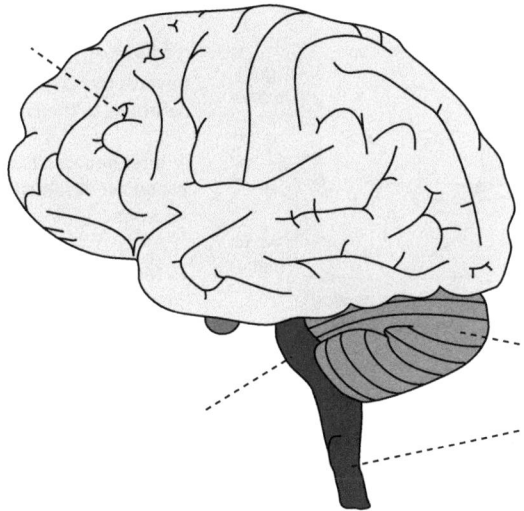

Cerebellum, Cerebrum, Medulla, Spinal cord

Sense organs have receptors that detect sensory data called stimuli. These stimuli are pieces of information about the bodies internal and external environment. This information is sent to the CNS via sensory neurons and the body's response to it is sent via motor neurons to effectors. Sometimes our body will need a fast, involuntary and automatic response to protect us from stimuli that may cause harm or damage. This type of response is known as a reflex action.

Reflex actions are not something we can learn, like riding a bike or playing a sport, or something we deliberately think about doing with our conscious mind.

Stimulus	Effect	Purpose
Touching the eye or dirt/dust entering the eye space.	Blinking rapidly, watery eyes	Clears potential harmful material away from the centre of the eye
Bright light	Pupils become smaller in bright light	High light intensity that can damage the cells at the back of the inner eye are protected
Dirt/pathogens entering the trachea (windpipe)	coughing	Clears the windpipe from potential harmful bacteria and pathogens

During a reflex action, the electrical impulses are sent through a pathway that can be shown in a diagram called a **Reflex arc**.

Example of a reflex arc:

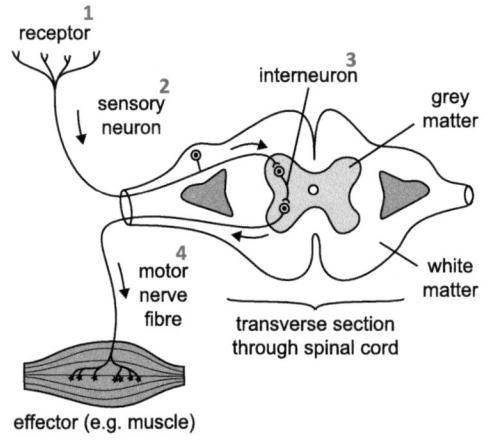

transverse section through spinal cord

Reflex arc

1. Pain receptors in the skin detect a harmful stimuli — sharp object example.

2. The electrical impulses are sent towards the CNS along sensory neurones to the inter neurons.

3. Inter neurons relay on protective response to the motor neurons.

4. Motor neurons send electrical impulses to the corrcet muscles (effectors) to cause them to contract. This contractions ensures the hand quickly pulls away from the pain stimuli before any damage, or further damage, can be done.

2.1.3 Practical: Reflex actions and the reflex arc

Practical: Reflex actions and the reflex arc

The process for reflex actions within an organism can be shown as:

Stimulus → Receptor → Sensory neuron → Inter neuron → Motor neuron → Effector → Response

One simple way to test reaction time in class or at home is by measuring the time it takes to catch a ruler dropped by a lab partner or friend.

Design an experiment to use the ruler-drop method to test reaction time.

Think about:

1. What will your aim and hypothesis be?
2. What apparatus will you need?
3. Do you have a step by step clear method or procedure written down?
4. What factors will you need to keep the same throughout the practical?
5. How will you record your results and how many results will you need to achieve a reliable and valid conclusion?
6. Can you think of how to take your practical further? What kind of things do you think would affect reactions time and how could you test this idea?

2.1.4 Nervous control: Questions

Nervous control: Questions   Go online

Q2: Describe the role of sensory and motor neurons.

Q3: Name the two parts of the Central Nervous System (CNS) in humans.

....

Q4: The following structures are involved in the transmission of nerve impulses. Arrange them in the correct order: Motor neuron, Sense organ, Muscle, CNS, Sensory neuron.

....

Q5: Choose the correct word from the list to complete the sentences:

1. The (cerebrum/ cerebellum/ medulla) is the section of the brain that controls memory, conscious thoughts, intelligence and emotions.
2. The (cerebrum/ cerebellum/ medulla) is the section of the brain that controls coordination, movements and balance.
3. The (cerebrum/ cerebellum/ medulla) is the section of the brain that controls breathing and heart rate.

2.2 Hormonal control

The endocrine glands release hormones into the bloodstream. Hormones are chemical messengers that transfer information from part of the body to another. Hormones can be carried in the blood to all the body organs but will only have an effect on their target organs. Different hormones are produced in different glands of the endocrine system.

TOPIC 2. CONTROL AND COMMUNICATION

Interactivity on the endocrine system Go online

Q6: Complete the diagram by the labels from the wordlist:

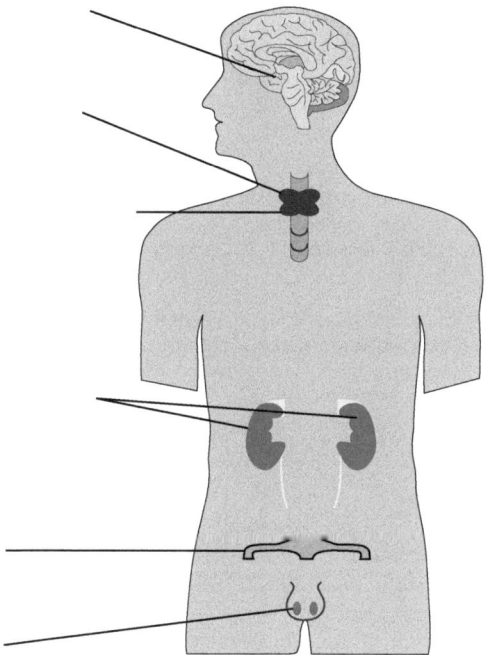

Ovary, Testis, Pituitary gland, Adrenal glands, Thyroid gland, Parathyroid gland

Target organs have tissues with cells containing complementary receptor proteins for specific hormones, so only that tissue will be affected by these hormones. As hormones move throughout the body, and come into contact with lots of different cells, it will only interact with a cell that has specific membrane receptors.

2.2.1 Target cell action

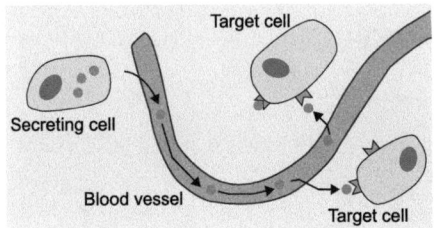
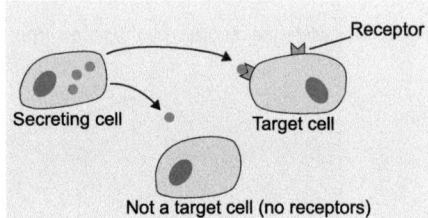

A secreting cell carrying a hormone that transfers into the blood stream for transportation to the target cell. As the hormone touches the target cells that have the correct receptors, it binds to the receptors.

In comparison, if the hormone touches a cell that does not have a matching receptor, nothing binding can take place.

In National 5 Biology, you do not need to know any of the names and locations of the individual endocrine glands other than those mentioned in the course as part of the learning content. The following table has two examples; the role of the pancreas and insulin is part of the course and the pituitary gland/ADH is extension information.

Gland	Target organ	Hormone	Outcome
Pancreas	Liver	Insulin	Controls blood glucose concentration
Pituitary gland	Kidney	ADH (anti-diuretic hormone)	Controls blood water concentration

2.2.2 Summary animation: Endocrine system animation

Watch the short video about endocrine system animation https://www.youtube.com/watch?v=IAXv8Kxt4QU .

The pancreas is the organ that produces digestive enzymes and the hormones **glucagon** and **insulin**. The pancreas is also known as an endocrine organ because it releases insulin and glucagon directly into the bloodstream to keep blood glucose levels regulated. The pancreas has receptor cells that detect any changes in blood glucose levels and responds by releasing the appropriate hormone which will then travel to the target organ, the **liver**.

The liver is the largest organ in the body and is involved in blood glucose control along with the hormones insulin and glucagon. The liver can store carbohydrates by removing excess glucose from the blood when the concentrations are too high, or release glucose back into the bloodstream when concentrations are too low.

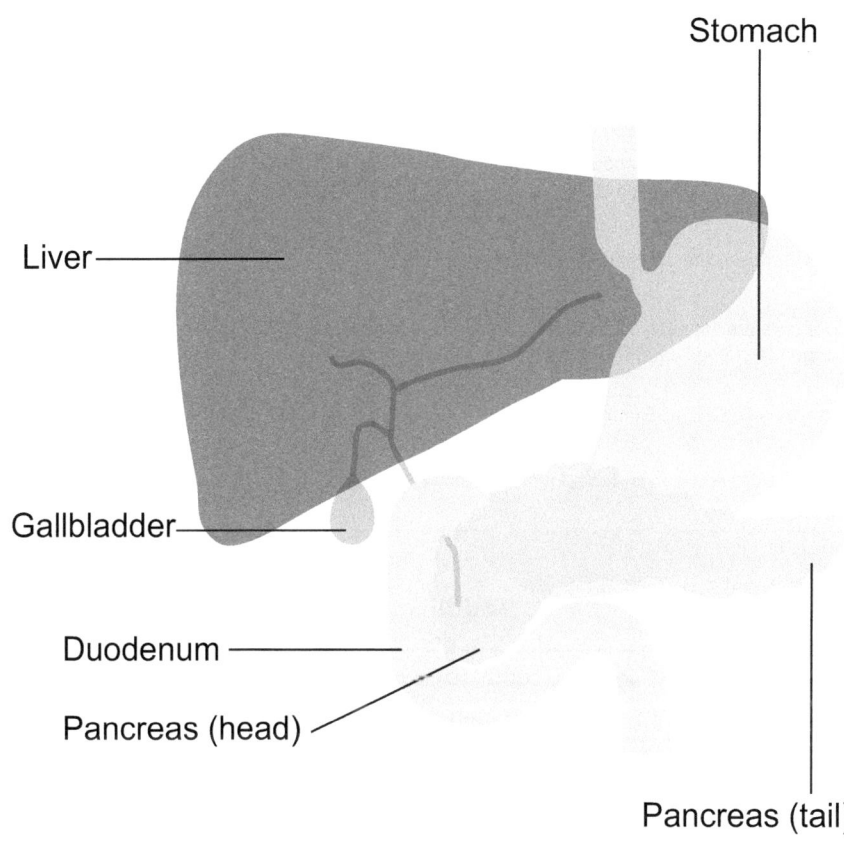

Stomach

Liver

Gallbladder

Duodenum

Pancreas (head)

Pancreas (tail)

Insulin is a hormone produced by the **pancreas** which triggers glucose conversion into glycogen, a storage carbohydrate. Glucagon is a hormone produced by the pancreas which triggers glycogen conversion into glucose. The following flow chart compares how the pancreas responds to changes in blood glucose level to maintain a steady glucose internal environment.

Increase in blood glucose levels	Decrease in blood glucose levels
Receptors in the pancreas detect the INCREASE in glucose levels	Receptors in the pancreas detect the DECREASE in glucose levels
↓	↓
The pancreas responds by producing the hormone insulin	Receptors in the pancreas detect the DECREASE in glucose levels
↓	↓
Insulin is carried in the bloodstream to the liver	Glucagon is carried in the bloodstream to the liver
↓	↓
The liver cells are stimulated to convert glucose to glycogen (a storage carbohydrate)	The liver cells are stimulated to convert glycogen to glucose.
↓	↓
This causes a DECREASE in blood glucose levels.	This causes an INCREASE in blood glucose levels.

A summary word equation for the hormonal action in this process is:

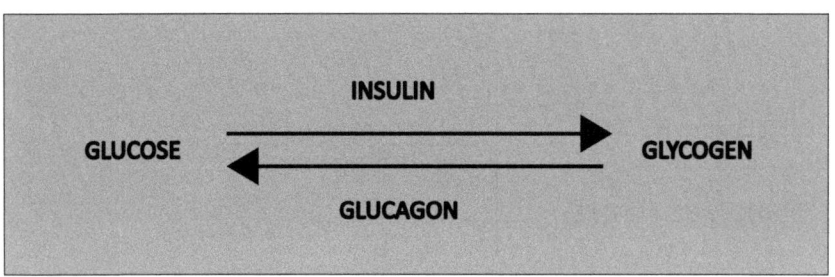

2.2.3 Summary animation: Brief overview of parts of the endocrine system and diabetes

Watch this short video about the endocrine system and diabetes. https://youtu.be/Ry5fTZfZHIs

A pictoral version of the information about maintaining the blood glucose concentration is shown below.

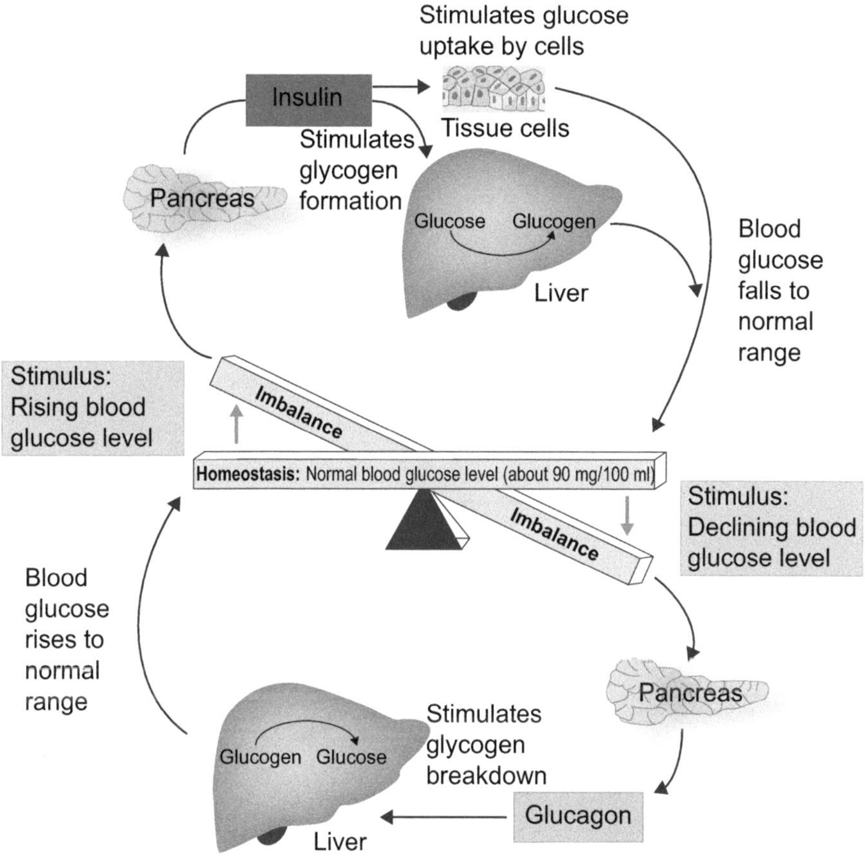

TOPIC 2. CONTROL AND COMMUNICATION

[Diagram showing blood glucose regulation: Rising blood glucose level stimulates the pancreas to release insulin, which stimulates glucose uptake by tissue cells and stimulates glycogen formation in the liver (Glucose → Glucogen), causing blood glucose to fall to normal range. Homeostasis is maintained at normal blood glucose level (about 90 mg/100 ml). Declining blood glucose level stimulates the pancreas to release glucagon, which stimulates glycogen breakdown in the liver (Glucogen → Glucose), causing blood glucose to rise to normal range.]

Diabetes is a disease that is caused by either a failure to release insulin (Type 1 Diabetes) or a failure to respond normally to insulin (Type 2 Diabetes).

The topic of Diabetes is discussed in detail in other parts of the National 5 course.

Research task: Causes and treatments of diabetes Go online

Investigate the causes and treatments of both Type 1 and Type 2 diabetes.

Starter questions:

1. What are the main differences between Type 1 and Type 2 diabetes?

2. What are the main causes of Type 1 and Type 2 diabetes?

3. Are there any factors lifestyle choices that increase the chances of diabetes? What are

© HERIOT-WATT UNIVERSITY

these and how could this be avoided?

4. Are there trends in Scottish health statistics to show and increase or decrease in the number of diabetics in the country? How does compare to the rest of the UK?

Hormonal Control: Questions　　　　　　　　　　　　　　　Go online

Q7: Describe the role of hormones in the body.

....

Q8: State the name of the glands that release hormones?

....

Q9: Which organ detects the changes in blood glucose concentrations?

Q10: Write a word equation to show the role of **insulin** in controlling blood glucose levels. Wordlist: Insulin, Glucose, Glycogen.

[ ]　　⟶　　[ ]

[ ]

....

Q11: Write a word equation to show the role of **glucagon** in controlling blood glucose levels. Wordlist: Glucose, Glucagon, Glycogen.

[ ]　　⟶　　[ ]

[ ]

2.3 Learning points

Summary

a) Nervous control

- The nervous system consists of central nervous system (CNS) and other nerves.
- The CNS consists of brain and spinal cord.
- The brain has structures called the cerebrum, cerebellum and medulla.
- The cerebrum is the section of the brain that controls memory, conscious thoughts, intelligence and emotions.
- The cerebellum is the section of the brain that controls coordination, movements and balance.
- The medulla is the section of the brain that controls breathing and heart rate.
- There are three types of neuron (sensory, inter and motor).
- Sensory neurons are nerve cells that carry electrical impulses from sense organs to CNS.
- Inter neurons are nerve cells that are found in the CNS where they connect with other neurons.
- Motor neurons are nerve cells that carry electrical impulses from the CNS to muscles and glands (effectors).
- Receptors detect sensory input/stimuli.
- Electrical impulses carry messages along neurons and chemicals transfer these messages between neurons, at synapses.
- Reflexes protect the body from harm.

b) Hormonal control

- The endocrine glands release hormones into the bloodstream.
- Hormones are chemical messengers
- Target tissues have cells with complementary receptor proteins for specific hormones, so only that tissue will be affected by these hormones.
- Insulin is a hormone produced by the pancreas which triggers glucose conversion into glycogen.
- Glucagon is a hormone produced by the pancreas which triggers glycogen conversion into glucose.
- Glycogen is stored carbohydrates.
- The pancreas is the organ that produces digestive enzymes and the hormones glucagon and insulin.
- The liver is the large organ involved in blood glucose control.

2.4 End of topic test

End of topic test: Control and communication   Go online

1. A footballer suffered a head injury during the game and began to lose their balance and show signs of dizziness.

Q12: Name the section of the brain that controls balance.

....

Q13: The doctor was worried that the footballer may have done some damage to their spinal chord and carried out a test to determine how sensitive the players skin was to a blunt needle. Describe how a stimulus, like a blunt needle, can be detected by the body.

- The electrical impulses are sent towards the spinal chord along sensory neurons to the inter neurones.
- Subject 'feels' the stimuli/ senses 'where' the needle is.
- Receptors in the skin detect the stimuli.

2. The following diagram shows a reflex response from a small animal.

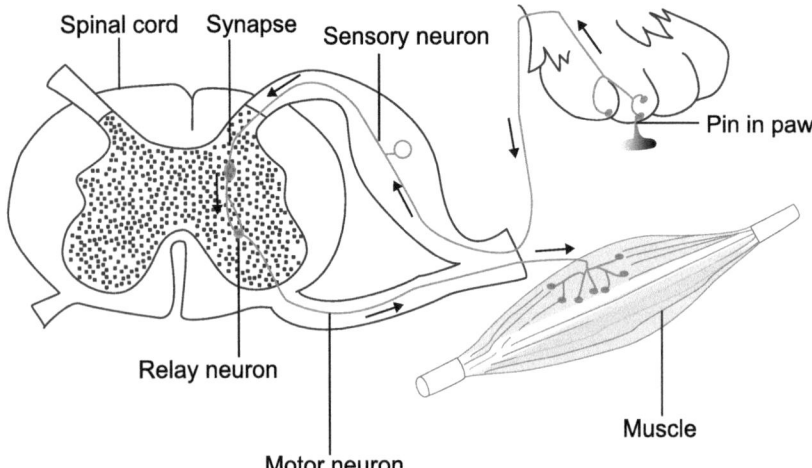

Q14: State the role of the sensory and motor neurons.

....

Q15: Complete the pathway to show how information is passed through a reflex arc.

Stimulus → ? → ? neuron → Inter neuron → ? neuron → Effector → Response

....

Q16: Identify what the a) stimulus is and b) effector are, in the diagram.

3. The diagram below shows a hormone binding to a cell in its target organ.

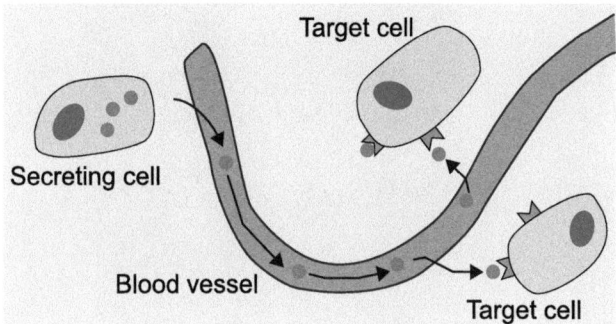

Q17: Explain why the target cells will be the only cells to be affected by this particular hormone?

....

Q18: Which type of gland will release the hormone into the blood stream so that it can travel to this target tissue?

Q19: Hormonal messages travel more slowly than nerve messages. State one other difference between these messages.

....

Q20:

Choose the correct word to complete the sentences:

Receptors in the **(pancreas / liver)** detect the increase in glucose levels and respond by producing the hormone **(insulin / glucagon)**. This results in the conversion of **(glucose / glycogen)** to **(glucose / glycogen)**.

Unit 2 Topic 3

Reproduction

Contents

3.1 Chromosomes and gametes 110
3.2 Reproduction in flowering plants 110
3.3 Reproduction in animals 114
3.4 Fertilisation 116
    3.4.1 Extension: Pollen tube formation 117
3.5 Learning points 119
3.6 End of topic test 119

Learning objective

At the end of this topic you should be able to:

- State that body cells are diploid and gametes (sex cells) are haploid.
- Name the types of gametes, the organs that produce them and where these are located in plants and animals.
- Describe the basic structure of sperm and egg cells.
- State that fertilisation is the fusion of the nuclei of the two haploid gametes to produce a diploid zygote, which divides to form an embryo.

3.1 Chromosomes and gametes

Most multicellular organisms contain cells that are have a diploid chromosome complement. This means that each cell contains two matching sets of chromosomes.

Multicellular organisms will produce sex cells to go through reproduction. These sex cells are known as gametes and have a haploid chromosome compliment. This means that each cell contains one set of chromosomes.

Gametes can be either male or female and, in both plants and animals, the fusion of the haploid gametes occurs with **fertilisation**.

3.2 Reproduction in flowering plants

Plants can reproduce new offspring by either asexual or sexual reproduction. Many flowering plants will use sexual reproduction to produce male and female **gametes** from the same flower.

The diagram below shows the main structures common to most flowering plants.

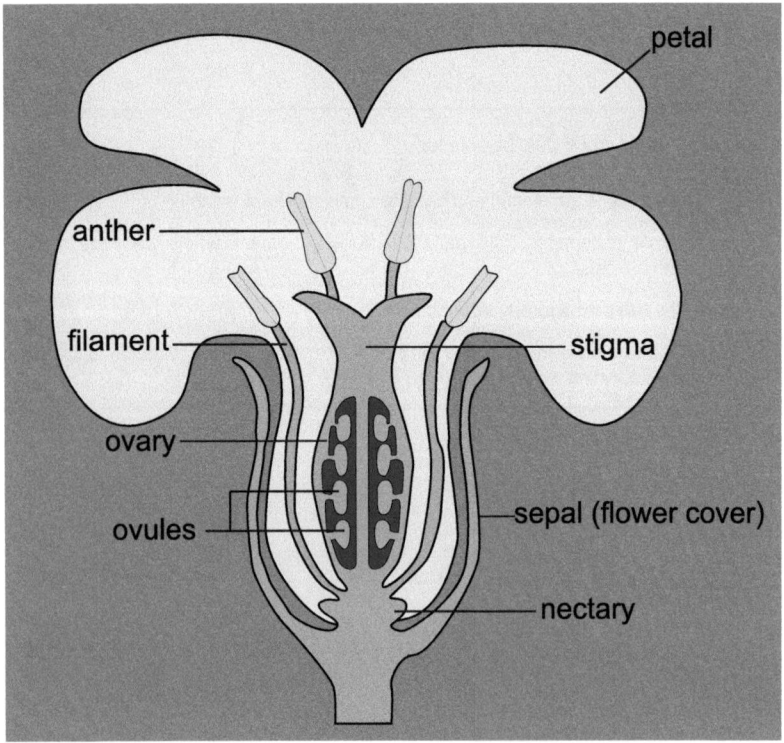

TOPIC 3. REPRODUCTION

Term	Definition
Anther	Organ in the flower that produces pollen grains.
Pollen grains	Structure containing the male gamete produced from the anthers of flowers.
Stigma	The top of the female part of the flower which collects pollen grains.
Ovary	Female organ in a flowering plant.
Ovules	Structure containing the female gamete produced from the ovaries of plants.

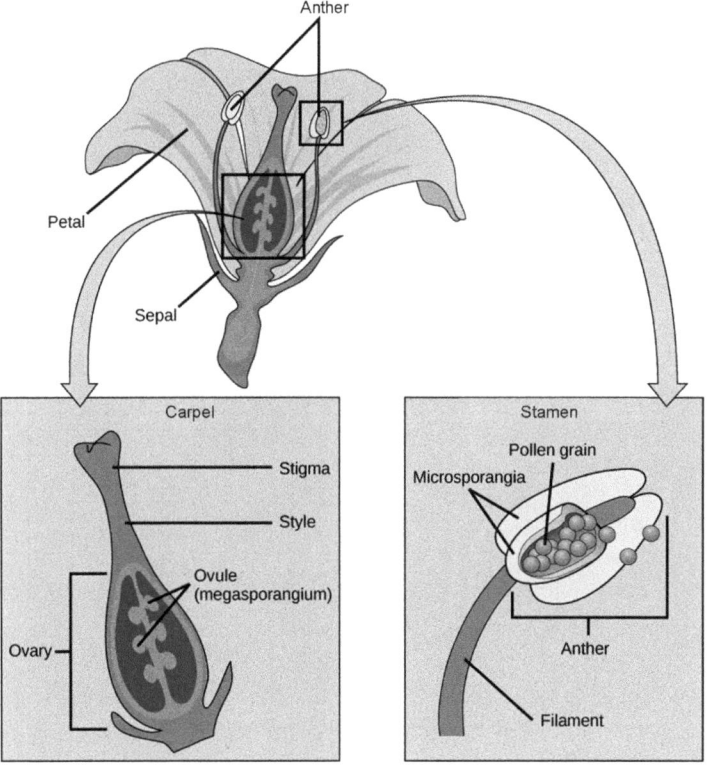

Male parts of flowering plants

The stamens are the male parts of flowering plants. They have an anther at the top and a filament stalk section. The male gametes are called pollen and are produced from the anther section. The picture below shows an anther with pollen grains containing the male gametes.

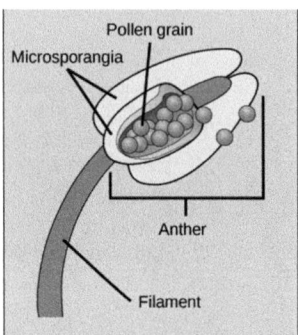

Male parts of flowering plants

Photograph - Male parts of flowering plants

Female parts of flowering plants

The stigma and ovary are the female parts of flowering plants. The male gamete, pollen, lands on the sticky stigma during **pollination** and a pollen tube will then grow down into the female ovary. The female gametes are called ovules and are produced in the ovary.

TOPIC 3. REPRODUCTION

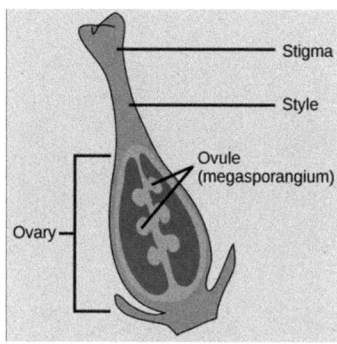

Female parts of flowering plants

Video: Pollination and fertilisation Go online

Watch this video about pollination and fertilisation.

https://youtu.be/aXT1DZEHsMk

Flower structure: Quiz Go online

Q1: Pick words from the word list below and add them into the right place:

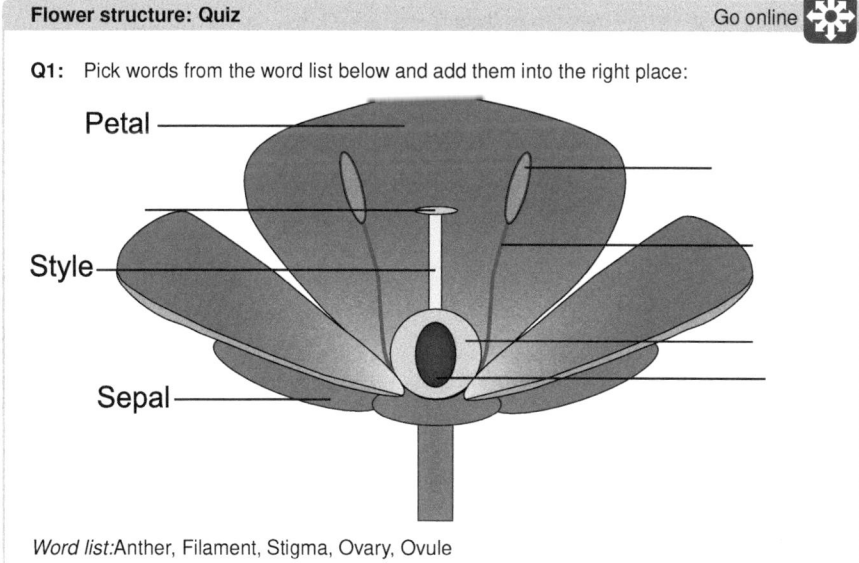

Word list: Anther, Filament, Stigma, Ovary, Ovule

114 UNIT 2. MULTICELLULAR ORGANISMS

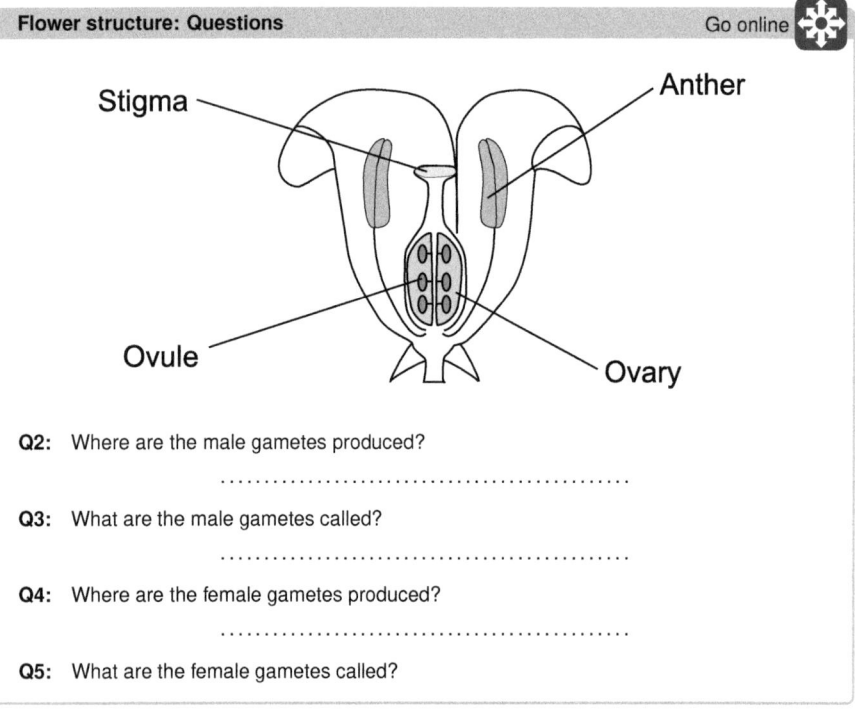

Flower structure: Questions   Go online

Q2: Where are the male gametes produced?

....

Q3: What are the male gametes called?

....

Q4: Where are the female gametes produced?

....

Q5: What are the female gametes called?

3.3 Reproduction in animals

Gametes in animals are produced in specialised organs called the reproductive organs.

Male reproductive system

The **teste** are the pair of organs within males that produce **sperm**. Sperm are the male gametes in animals. When the testes produce sperm, sperm travel along the sperm duct towards the penis and leaves the penis through the urethra.

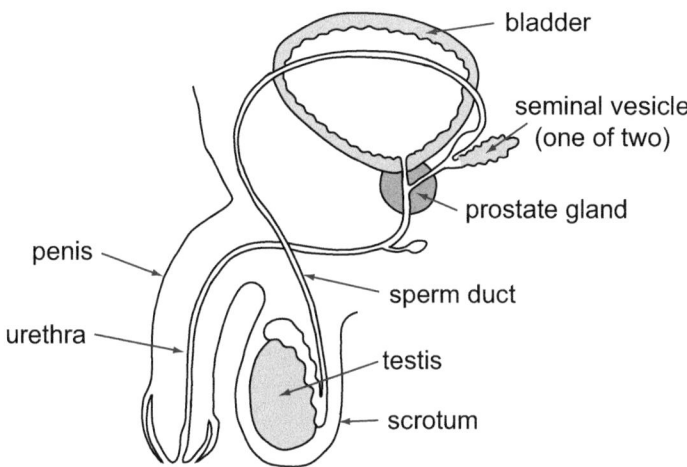

Female reproductive system

The ovaries are the organs within females that produce eggs. Eggs are the female gametes in animals. When the ovaries produce an egg, it will travel along the oviduct towards the uterus. If an egg is fertilised by a sperm as it travels along the oviduct, a **zygote** will be formed. A zygote is a fertilised egg cell that will go on to divide many times by the process of mitosis.

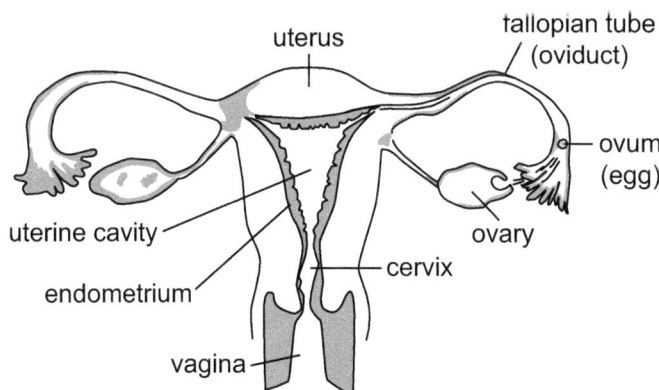

The following table shows a comparison of sperm and egg

Aspect	Sperm	Egg
Site of production	In the testes	In the ovaries
Size of cells	Small	Big
Cell structure	Has a head, middle section and tail	Round with a large food store in its cytoplasm
Number produced	Millions	One per ovulation cycle
Motility	Very mobile; ability to swim	Not mobile; helped to travel by cilia hairs in the oviduct

Video: Mitosis of a fertilised egg Go online

Watch this video about mitosis of a fertilised egg.
https://youtu.be/lsLZW0qUXKQ

Reproduction in animals: Questions Go online

Q6: Describe how sperm is suited to its function.
....

Q7: Describe how egg is suited to its function.

3.4 Fertilisation

Fertilisation is the fusion of the nuclei of the two haploid gametes to produce a diploid zygote. The male gamete nucleus fuses with the female gamete nucleus, both of which contain one set of chromosomal information each. The zygote that is formed at fertilisation receives each of these sets of chromosomal information to have two sets (diploid) and can then divide to form an embryo.

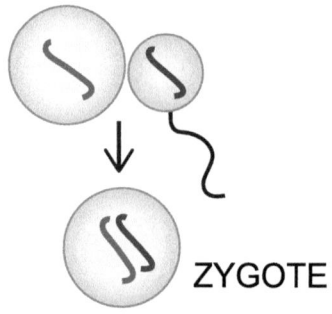

TOPIC 3. REPRODUCTION

Fertilisation: Quiz   Go online

Q8: Pair the terms from the wordlist with their correct definition.

Wordlist:*Gamete, Haploid, Zygote, Fertilisation, Diploid*

Term	Definition
	The fusion of two gametes
	Sex cell containing the haploid chromosome number
	A fertilised egg
	One set of chromosomes
	Two matching sets of chromosomes

Video: Fertilisation of a human egg (by IVF)   Go online

Watch this time lapse video about the fertilisation of a human egg (by IVF) under the microscope.

https://youtu.be/Z-hhTHe5FbQ

3.4.1 Extension: Pollen tube formation

Pollination is the transfer of pollen from the anther to the stigma. This can be from a one flower to another (cross-pollination) or from one flower to itself (self-pollination.)

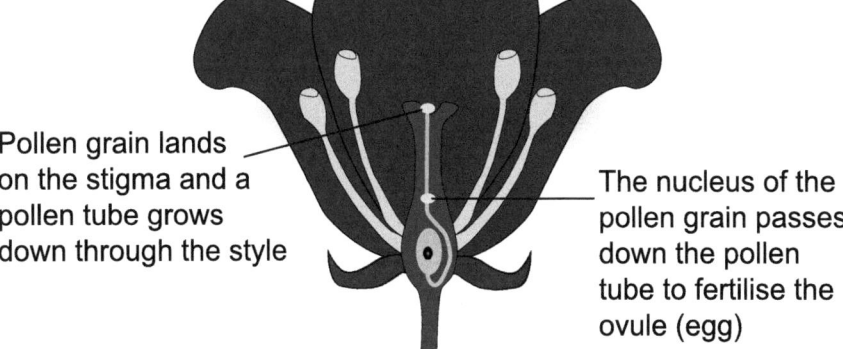

Pollen grain lands on the stigma and a pollen tube grows down through the style

The nucleus of the pollen grain passes down the pollen tube to fertilise the ovule (egg)

After pollination takes place, the pollen grain begins to grow a pollen tube.

118 UNIT 2. MULTICELLULAR ORGANISMS

Video: Pollen tube formation	Go online
Watch this time lapse video about pollen tube formation. https://youtu.be/Oppq2VmCJjM	

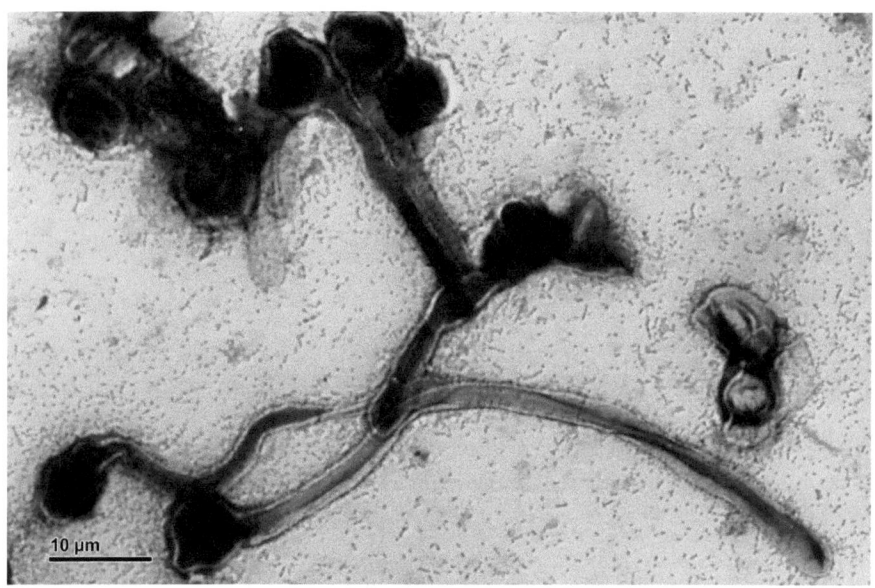

Growth of the pollen tube following pollination

Stage	Description
Stage 1	The pollen grain begins to grow a pollen tube through the tissues of the style and towards the ovary.
Stage 2	The nucleus inside the pollen grain starts to make its way down the inside of the tube.
Stage 3	When the end of the pollen tube reaches and ovule in the ovary, it enters via a tiny hole.
Stage 4	The tip of the pollen tube will then burst to release the male gamete so that it may fuse with the female gamete and fertilisation can take place.

© HERIOT-WATT UNIVERSITY

3.5 Learning points

Summary

- Body cells are diploid multicellular organisms and are haploid in gametes (sex cells).
- Diploid means that each cell contains two sets of matching chromosomes.
- Haploid means that each cell contains one set of chromosomes.
- Gametes are sex cells and can be either male or female.
- The organs that produce gametes in plants are anthers and ovaries. Anthers produce pollen grains containing the male gamete. Ovaries contain ovules which contain the female gamete.
- The organs that produce gametes in animals are testes and ovaries. Testes produce the male gamete, sperm. Ovaries produce the female gamete, ova (egg).
- Sperm are very small and motile. Ova are large because they contain a food store and can not move by themselves.
- Fertilisation is the fusion of the nuclei of the two haploid gametes to produce a diploid zygote, which divides to form an embryo.

3.6 End of topic test

End of topic test: Reproduction Go online

Q9: The diagram shows some of the structures in a flower. Which of the following is produced in the structure labelled z?

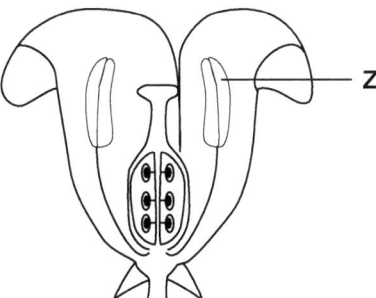

a) Pollen
b) Anther
c) Ovule
d) Ovary

Q10:
Match the terms in the table below to the correct definition

Term	Definition
Zygote	The fusion of two gametes
Diploid	Sex cell containing the haploid chromosome number
Haploid	A fertilised egg
Gamete	One set of chromosome
Fertilisation	Two matching sets of chromosomes

Q11:
Match the terms in the table below to the correct definition

Term	Definition
Ovary	Organ in the flower that produces pollen grains
Ovules	Structure containing the male gamete produced from the anthers of flowers
Pollen grains	The transfer of pollen grains from an anther to a stigma
Pollination	Female organ in a flowering plant
Anther	Structure containing the female gamete produced from the ovaries of plants

Unit 2 Topic 4

Variation and inheritance

Contents

4.1	Variation in species	122
4.2	Genetic terminology	128
4.3	Monohybrid crosses	128
4.4	Phenotypic ratios	134
4.5	Learning points	135
4.6	End of topic test	136

Learning objective

At the end of this topic you should be able to:

- Make comparisons between discrete variation (single gene inheritance) and continuous variation (polygenic inheritance).

- Give named examples of discrete variation and continuous variation.

- Give the definitions of genetic terms: gene; allele; phenotype; genotype; dominant; recessive; homozygous; heterozygous and P, F_1 and F_2.

- Carry out monohybrid crosses from parental generation through to F_2 generation.

- Use Punnett squares to explain inheritance.

- State the reasons why predicted phenotype ratios among offspring are not always achieved.

4.1 Variation in species

A species is a group of living things which have so many similarities that they can interbreed and produce fertile offspring. If the members of two different species were to breed, they would not be able to produce fertile offspring and, in most cases, cannot produce any offspring.

The members of the same species are not identical to one another, but will have many characteristics that are similar. This may be colour, size, shape etc. These differences are all due to inherited characteristics and the impact of the environment around the species. The term that is used to describe these inherited and environment differences on and in a species is '**variation**.'

Combining genes from two parents contributes to variation within a species. There are two ways to categorise the variations;

1. **Discrete** variation is the single gene inheritance of characteristics where measurements fall into distinct groups.

2. **Continuous** variation is the **polygenic** inheritance of characteristics where there is a range of values from one extreme to the other.

Video: What is variation?	Go online
Watch this video about variation in species. https://youtu.be/aG8fMxaSSNw	

Discrete variation

These are variations where the characteristics of the members of the species can be divided into distinct groups or categories.

TOPIC 4. VARIATION AND INHERITANCE   123

	Variation	Examples of variations
	Ear lobes	Attached or unattached
	Blood types	O, B, A or AB types

	Variation	Examples of variations
	Tongue rolling	Tongue roller or non-tongue roller
	Eye colour in fruit flies	White or red
Round Wrinkled	Pea coats	Round or wrinkled

The diagrams below show the type of graph that can be drawn for the two examples of discrete

© HERIOT-WATT UNIVERSITY

variation data. A bar graph or pie chart would be the most appropriate way to represent any data that has categories and distinct choices.

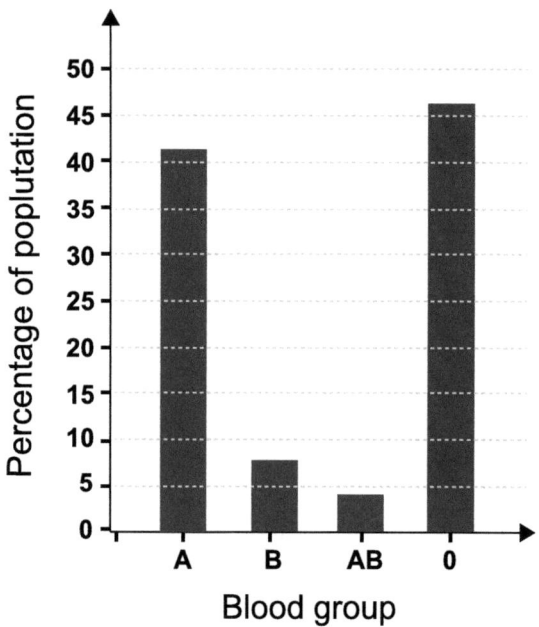

Blood group

Human blood groups (above graph) are a very good example of discrete variation. There are only four possible blood groups (O, A, B and AB) and no other values inbetween. A bar graph or pie chart is the best way to visually represent the data.

Tongue rolling (below graph) is another example of discrete variation represented by a bar graph.

Number of Tongue Rollers in Sample

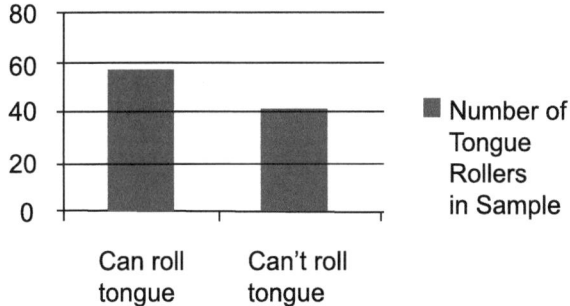

Continuous variation

The characteristics shown in discrete variation are controlled by single genes but the majority of characteristics shown by plants and animals are controlled by many genes acting together. These are variations, where the characteristics of the members of the species show a continuous range of possibilities between two extremes or a minimum and maximum value, are known as polygenic characteristics. There will be no categories or distinct groups that members can identify with, but instead they will be able to be placed somewhere in the data range. Examples could be weight, height and handspan. The diagram below shows the type of graph that can be drawn for the number of people at each height (cm). This graph is described as a bell-shaped curve and shows the normal distribution range.

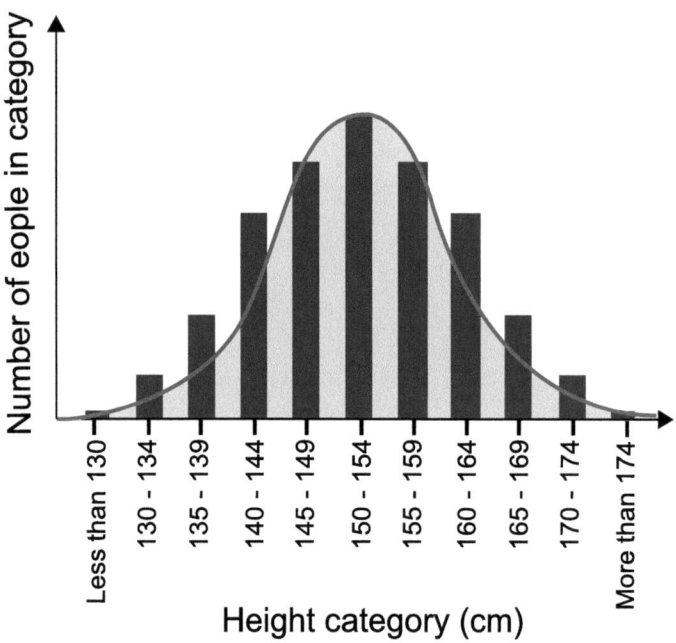

TOPIC 4. VARIATION AND INHERITANCE

Variation in species: Questions Go online

Q1: Match the following terms to their correct definition.

Term	Meaning
Continuous	Variations that are able to be categories into groups
Discrete	Variations that seen as one extreme to the other with a range of values in between
Variation	Differences in the characteristics of members of the same species

Q2: What is a species?

....

Q3: What is continuous variation?

....

Q4: What is discrete variation?

4.2 Genetic terminology

The following terms and abbreviations/symbols are commonly used in the study of genetics.

Term	Meaning
Gene	Each gene controls an inherited characteristic. A gene is the basic unit of inheritance and many genes together make up a chromosome.
Allele	The form of a gene. Different alleles (alternative forms of alleles) will result in variations of characteristics.
P	The symbol used to represent the parent generation in a cross.
F_1	'First filial generation'. This is the symbol used to represent the first generation in a cross.
F_2	'Second filial generation'. This is the symbol used to represent the second generation in a cross.
Genotype	The particular alleles that an organisms has for a genotype.
Heterozygous	Two different alleles of a genotype ie. Aa or Bb.
Homozygous	Two alleles the same for a genotype ie. AA or aa.
Phenotype	The physical appearance expressed by an organisms due to their genotype.
Dominant	The form of a gene which is always expressed.
Recessive	The form of a gene which will only be expressed if the genotype is homozygous.
Diploid	Two matching sets of chromosomes. The symbol for this is '2n'.
Haploid	One set of chromosomes. The symbol for this is 'n'.

4.3 Monohybrid crosses

A monohybrid cross is the model that is used to show the inheritance pattern of **one** characteristic. The genes that an organism has is made up of two alleles, one allele from each parent. Therefore, alleles are the different forms of the gene and can be described as being recessive or dominant.

Recessive: The form of a gene which will only be expressed if the genotype is homozygous. Normally this will be represented by a lowercase letter- **'a'** and would only be expressed as a phenotype if both alleles of the gene were recessive i.e. **'aa'**.

Dominant: The form of a gene which is always expressed and is normally shown as an uppercase letter- **'A'**. Even if a gene had two different alleles i.e. **'Aa'**, the dominant allele will always be expressed as it masks the recessive allele.

This information will be identified in a monohybrid cross as the term 'genotype.' The physical characteristic that is expressed by a genotype is referred to as a 'phenotype.' Remember the genotype always determines the phenotype.

TOPIC 4. VARIATION AND INHERITANCE

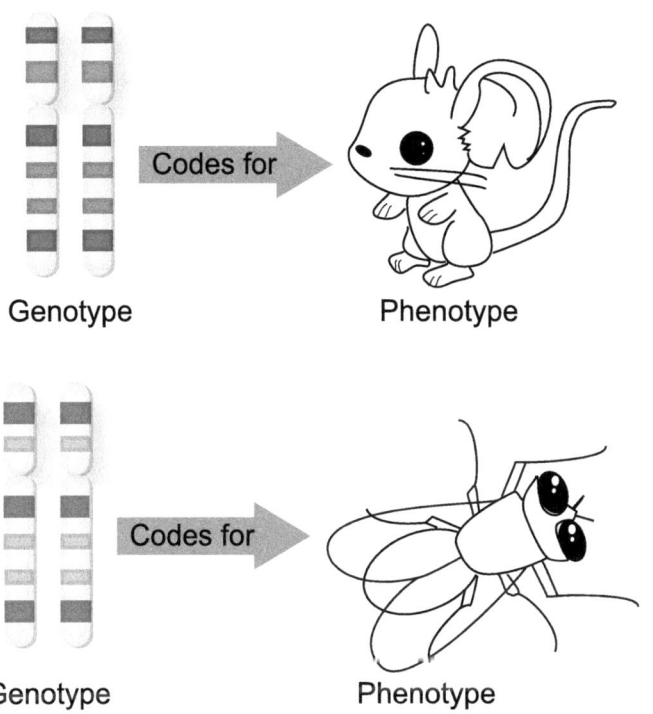

Video: How Mendel's pea plants helped us understand genetics Go online

Watch this TED-Ed talk about "How Mendel's pea plants helped us understand genetics".

https://youtu.be/Mehz7tCxjSE

In roses, red flower colour is dominant to white flower colour. Below is a table to show how different allele combinations result in different phenotypes being expressed.

Allele 1	Allele 2	Phenotype expressed
Red flower	Red flower	Red flower
Red flower	White flower	Red flower
White flower	White flower	White flower

The set of alleles that an organism has on one gene (its genotype) can be described using one of two words as shown in the table below.

Term	Description	Example of allele combination
heterozygous	Two different forms of the allele of a genotype	Aa or Bb
homozygous	Two alleles the same for a genotype	AA or aa and BB or bb

The diagram that follows is a monohybrid cross of flower colour in pea plants. It has been labelled to show the information in this topic so far.

The next diagram shows more detail on how we can predict the F2 generation genotype and phenotypes using the F1 information and a punnet square.

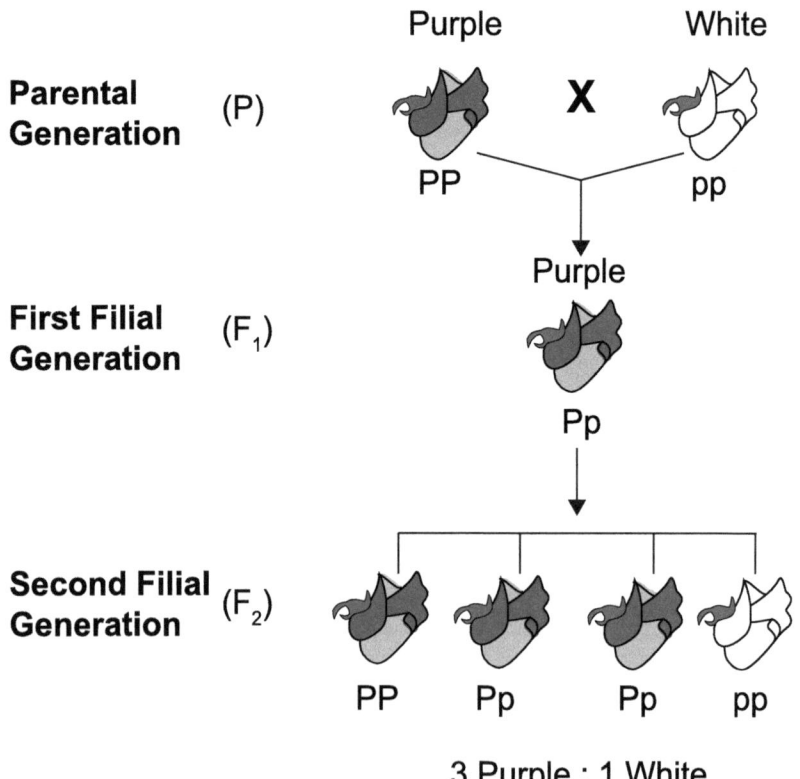

3 Purple : 1 White

- Flower colour is controlled by two different alleles — purple or white.
- Purple flower colour is the dominant allele (P) and white is the recessive allele (p).
- Both parents are homozygous (Purple = PP and white = pp).
- The F_1 generation are heterozygous (Pp) as they have inherited one allele from each parent.

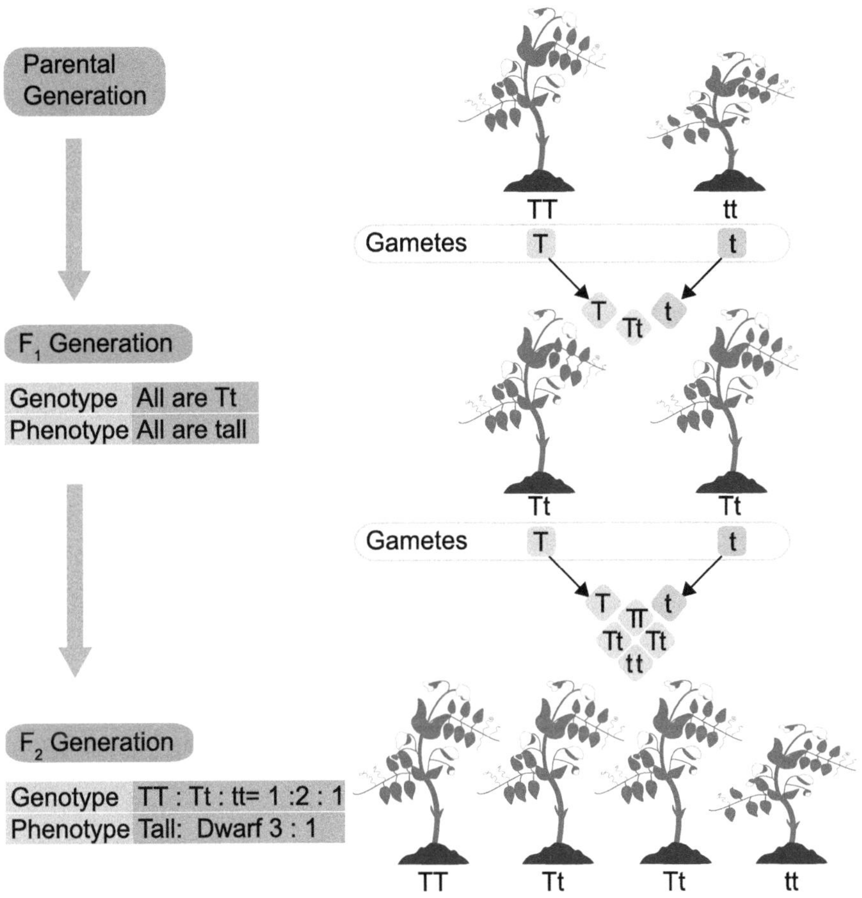

TOPIC 4. VARIATION AND INHERITANCE

Parental Generation

TT × tt
Gametes: T, t

F₁ Generation

Genotype: All are Tt
Phenotype: All are tall

Tt × Tt
Gametes: T, t

F₂ Generation

Genotype TT : Tt : tt = 1 : 2 : 1
Phenotype Tall : Dwarf 3 : 1

TT, Tt, Tt, tt

- Height is controlled by two different alleles — tall or dwarf.
- Tall height is the dominant allele (T) and dwarf is the recessive allele (t).
- Both parents are homozygous (Tall = TT and dwarf = tt).
- The F₁ generation are heterozygous (all Tt) as they have inherited one allele from each parent.
- One F₁ parent gamete possibilities (T or t) are placed along one side of the punnet and the other parent gamete possibilities (T or t) are placed along the other side.

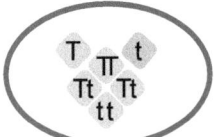

© HERIOT-WATT UNIVERSITY

- The punnet square lets us fill in all the different possible combinations for the F_2 generation.

| Video: Genetic diagrams | Go online |

Watch this video from the GCSE Biology Genetic diagrams (Edexcel 9-1).

https://youtu.be/5Vmll0bRf_A

| Monohybrid crosses: Questions | Go online |

Q5: What are chromosomes?

....

Q6: How many chromosomes are present in the nucleus of a normal human body cells?

Q7:
Choose the correct word from the drop down menu to complete the sentences.
During gamete formation, **(sex/body)** cells receive different combinations of the **(paired/single)** chromosomes present in the original gamete mother cell.

....

Q8:
Match the following terms to their correct definition.

Term	Meaning
Polygenic	The form of a gene
Phenotype	The particular alleles that an organisms has for a genotype
Homozygous	Two different alleles of a genotype i.e. Aa or Bb
Genotype	Two alleles the same for a genotype i.e. AA or aa
Recessive	The physical appearance expressed by an organisms due to their genotype
Heterozygous	Type of inheritance involving several genes acting together
Allele	The form of a gene which will only be expressed if the genotype is homozygous

| Interactivity: Punnet squares | Go online |

As shown previously, punnett squares can be used to predict the results of any cross where the genotypes of the parents are known.

TOPIC 4. VARIATION AND INHERITANCE

Use the information below to practice completing punnet squares.

Q9: Predict the genotypes and phenotypes of cat offspring tail length from a cross between a long tailed cat (HH) and a short tailed cat (hh).

Parental genotypes	h	h
H		
H		

Q10:

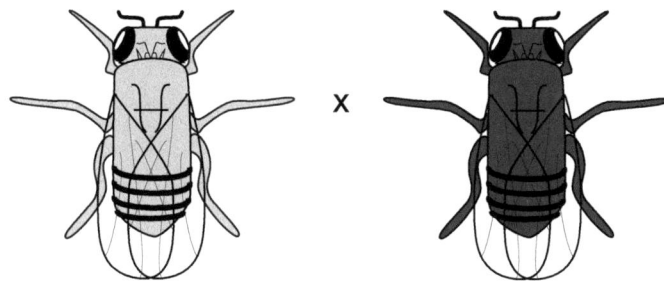

Genotype: GG   gg
Phenotype: Grey body   Black body

Body colour in fruit flies is an example of _____ variation.

a) discrete
b) continuous

....

Q11: From the parent genotypes shown in the diagram, the F_1 flies produced will be _____ .

a) homozygous
b) heterozygous

....

Q12: Complete the punnet square to show your answer to the F_1 genotypes.

Parental genotypes	g	g
G		
G		

© HERIOT-WATT UNIVERSITY

4.4 Phenotypic ratios

Predicted phenotype ratios among offspring are not always achieved because of the random nature of fertilisation and the fusion of the genetic information in gametes.

Interactivity: Phenotypic ratios   Go online

In a survey of 90 students it was found that 25 of them had hitchhiker's thumb.

Q13: Calculate the number of students with straight thumb to hitchhiker's thumb as a simple, whole number ratio.

☐

:

☐

....

Q14: The predicted ratio was 3 straight thumb : 1 hitchhiker's thumb. Explain why the predicted ratio was different to the actual ratio.

4.5 Learning points

Summary

- Combining genes from two parents contributes to variation within a species.
- Discrete variation is the single gene inheritance of characteristics where measurements fall into distinct groups.
- Continuous variation is the polygenic inheritance of characteristics where there is a range of values from one extreme to the other.
- Genetic terms such as gene; allele; phenotype; genotype; dominant; recessive; homozygous; heterozygous and P, F_1 and F_2 help the understanding of how genetic information is passed from parents to offspring.
- Monohybrid crosses is the model that is used to show the inheritance pattern of one characteristic.
- Punnett squares can be used to predict the results of any cross where the genotypes of the parents are known.
- Predicted phenotype ratios among offspring are not always achieved because of the random nature of fertilisation and the fusion of the genetic information in gametes.

4.6 End of topic test

End of topic test: Variation and inheritance Go online

Q15: Most features of an individual phenotype are:

a) controlled by a single gene and show continuous variation
b) controlled by a single gene and show discrete variation
c) polygenic and show continuous variation
d) polygenic and show discrete variation.

....

The following diagram shows the inheritance of coat colour in guinea pigs.

P	Phenotype:	Black guinea pig X White guinea pig
P	Genotype:	BB bb
F_1	Genotype:	Bb
F_2	Genotypes:	BB and Bb and bb

Q16: Which of the following generations contain heterozygous individuals?

a) P and F_1
b) P and F_2
c) F_1 and F_2
d) P, F_1 and F_2

The diagrams below show the same sections of matching chromosomes found in four flies, A, B, C and D.

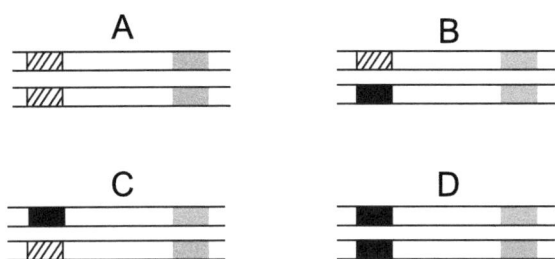

The alleles shown on the chromosomes can be indentified using the following key.

▨ allele for striped body
■ allele for unstriped body
▢ allele for normal antennae
▢ allele for abnormal antennae

Which fly is homozygous for body pattern and heterozygous for antennae type?

Q17: Which fly is homozygous for body pattern and heterozygous for antennae type?

....

Q18: Most features of an individual's phenotype are controlled by more than one gene. Name this type of inheritance.

....

Q19:
Variation in a characteristic can either be discrete or continuous. The range of heights and weights for a group of students were measured and recorded. Ear lobe types were also examined and categorised into groups.
Which line in the table below identifies the type of variation shown by each of these human characteristics?

	Height	Weight	Ear lobe types
A	continuous	continuous	discrete
B	discrete	continuous	continuous
C	discrete	discrete	continuous
D	continuous	discrete	discrete

© HERIOT-WATT UNIVERSITY

Unit 2 Topic 5

Transport systems of plants

Contents

5.1 Plant organs 140
5.2 Water transport and the xylem 142
5.3 The process of transpiration 143
5.4 Sugar transportation and the phloem 145
5.5 Extension work: Comparison of xylem and phloem vessels in plants 146
5.6 Learning points 149
5.7 End of topic test 150

Learning objective

At the end of this topic you should be able to:

- State that plant organs are roots, stems and leaves.
- Label a leaf structure diagram showing upper epidermis, palisade mesophyll, spongy mesophyll, vein (consisting of xylem and phloem), lower epidermis, guard cells and stomata.
- State the function of each part labelled in a leaf structure diagram.
- State that water and minerals enter the plant through the root hairs and are transported in dead xylem vessels.
- Describe the structure of xylem vessels.
- Give a definition of the process of transpiration.
- Describe how the rate of transpiration is affected by wind speed, humidity, temperature and surface area.
- State that sugar is transported up and down the plant in living phloem.
- Describe the structure of phloem tissue.

5.1 Plant organs

All living things require water and sugar to stay alive. Plants have roots buried into the ground to allow them to absorb water and green leaves above the ground to help them to make sugar by the process of photosynthesis. This means that a plant needs a good transport system so that water can travel up to leaf cells and sugar can pass to many different locations across the plant where it is needed.

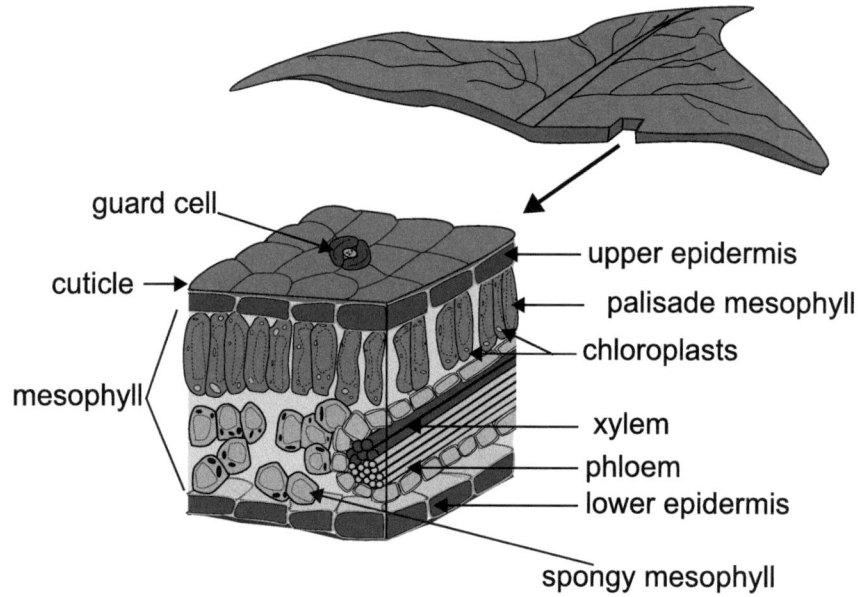

TOPIC 5. TRANSPORT SYSTEMS OF PLANTS

Structure	Function
Guard cells	Control the opening and closing of stoma and when closed can prevent water loss.
Upper epidermis	This is a single layer of cells containing few or no chloroplasts. The cells are fairly transparent and allow most of the light that strikes them to pass through to the cells below them.
Lower epidermis	The lower epidermis contains stomata cells that help prevent water loss and regulate the exchange of gases such as oxygen and carbon dioxide.
Phloem	Vessel in plants that transports sugars.
Spongey mesophyll	Loosely packed plant leaf tissue with air spaces for gas exchange.
Stomata	Tiny pores that allow for gas exchange in the leaf epidermis.
Palisade mesophyll	Contain the most chloroplasts to carry out most of photosynthesis.
Xylem	Narrow, dead tubes with **lignin** walls that transports water and minerals in plants
Waxy cuticle	A waxy layer to reduce water loss.

Stomata are tiny pores found on the surface of a leaf. Plants take in carbon dioxide from the air through stomata and water vapour will pass out.

Microscopic images of stomata link   Go online

Have a look at images of stomata under the microscope https://bit.ly/2BWdK8k.

Video: Leaf structure   Go online

Watch this video about leaf structure.
https://youtu.be/co0JdqUlycg

Interactivity: Leaf structure

Go online

Q1:

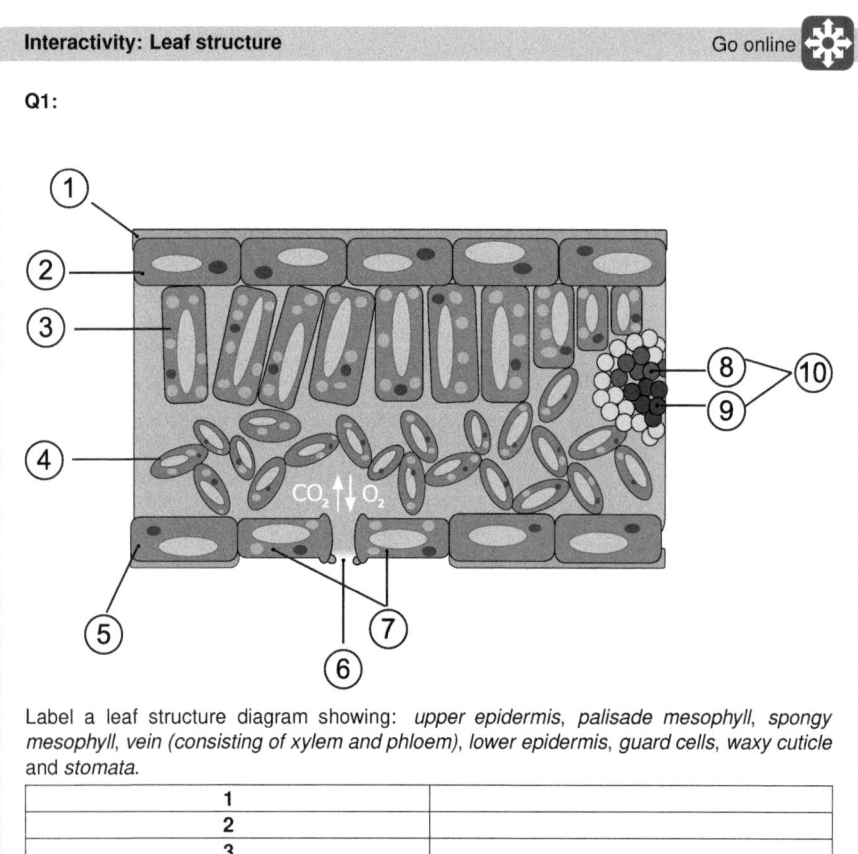

Label a leaf structure diagram showing: *upper epidermis, palisade mesophyll, spongy mesophyll, vein (consisting of xylem and phloem), lower epidermis, guard cells, waxy cuticle* and *stomata*.

1	
2	
3	
4	
5	
6	
7	
8	
9	
10	

5.2 Water transport and the xylem

The movement of water in a plant is in an upwards direction only. Water enters the plant by the process of osmosis (from a high water concentration outside the plant in the soil to a low water concentration inside the plant root hair cells.) It will then move from cell to cell again by osmosis.

TOPIC 5. TRANSPORT SYSTEMS OF PLANTS

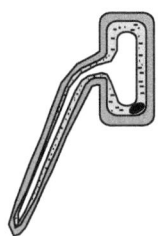

Plant root hair cells absorb the water and the water moves up the stem through the xylem vessels and into the leaf area.

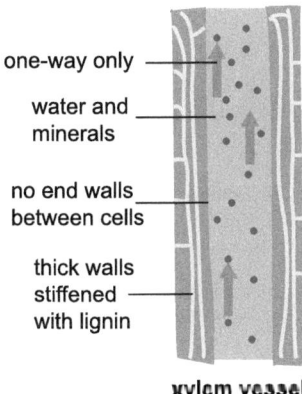

one-way only

water and minerals

no end walls between cells

thick walls stiffened with lignin

xylem vessel

Xylem vessels are formed from elongated cells which have died. The cells that are left become strengthened.

Lignin on the inside of the cell walls of xylem are shaped in spiral or ring patterns and this helps the xylem withstand the high pressure changes of water as it moves upwards through the plant.

5.3 The process of transpiration

The loss of water from plant leaves is known as **transpiration**. There is always a movement of water by osmosis in the leaves of a plant.

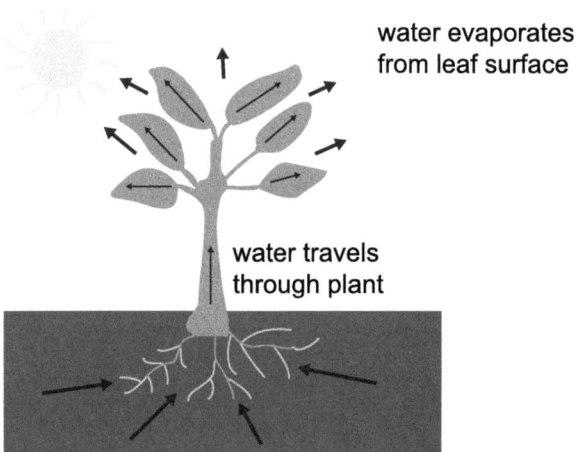

Water will evaporate into the air spaces from the mesophyll cells. Sometimes this will occur at different speeds depending on the environmental factors surrounding the plant. The stomata on the lower epidermis of the leaf are the final exit point of any water from the leaf.

The rate of transpiration is affected by abiotic factors such as wind speed, humidity and temperature. The diagram below shows a potometer which is one of the main pieces of apparatus to measure the rate of transpiration in plants.

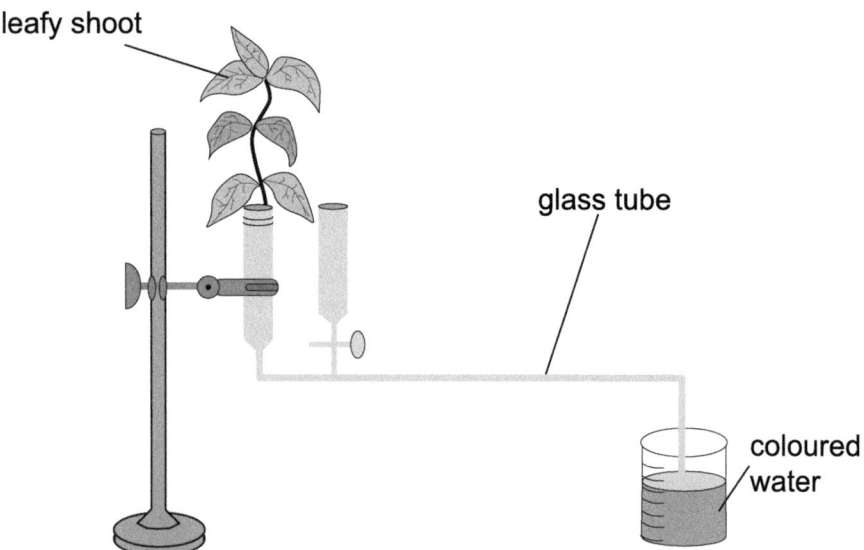

A potometer measures the rate of water absorption by plant shoots. If water is lost by transpiration, the upwards movement of water can be measured as it is drawn up from the water reservoir via the glass tube.

Changes in the surrounding environment can have an effect on the rate of transpiration. The environmental changes below all have an effect this change has on the rate of transpiration.

1. increase in humidity; Decreases the rate of transpiration
2. increase in temperature; Increases the rate of transpiration
3. increase in wind speed; Increases the rate of transpiration

5.4 Sugar transportation and the phloem

Sugar can be transported, after production by photosynthesis, in any direction across a plant. It can be used in a variety of ways.

1. Used immediately in respiration to release energy for use by other plant cells.
2. Converted into cellulose to be used in plant cell walls.
3. Converted to starch for storage until required.

Phloem cells

The transport of sugar takes place in vessels called the phloem. Phloem tissue is formed from elongated cells that are then arranged into vessels and then highly adapted to transport sugar more efficiently. Phloem vessels are still living, unlike the xylem vessels that are dead.

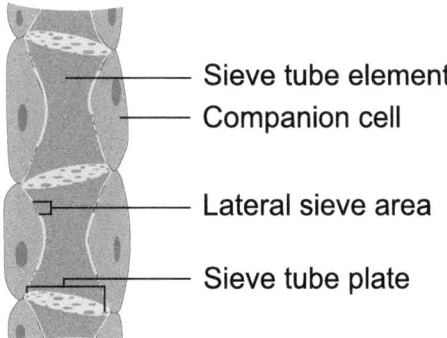

Sieve tube element
Companion cell
Lateral sieve area
Sieve tube plate

5.5 Extension work: Comparison of xylem and phloem vessels in plants

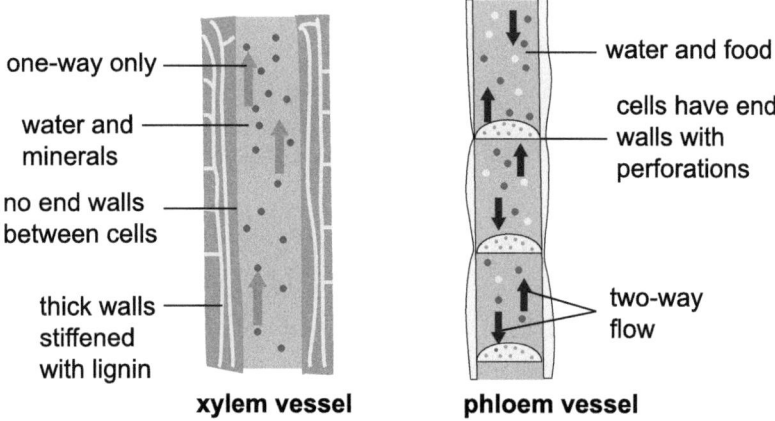

Xylem vessels make up a series of strong lignified tubes running from the root to the leaves and therefore help support the plant. In the two diagrams below, the xylem vessels can be seen in the more central area, helping to provide the upright structure of stems and tree trunks.

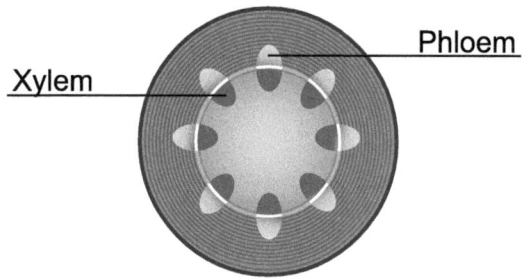

Cross Section of Stem and Root (below)

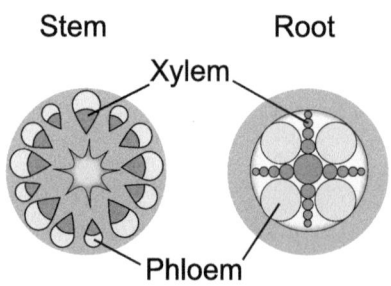

TOPIC 5. TRANSPORT SYSTEMS OF PLANTS

Interactivity: Estimating the age of a tree   Go online

In perennial plants, the xylem vessel develops in rings which can be most easily seen in trees. A lot of the wood of a tree is made up of xylem vessels and each year a new ring of xylem will develop on the outside of the tree trunk.

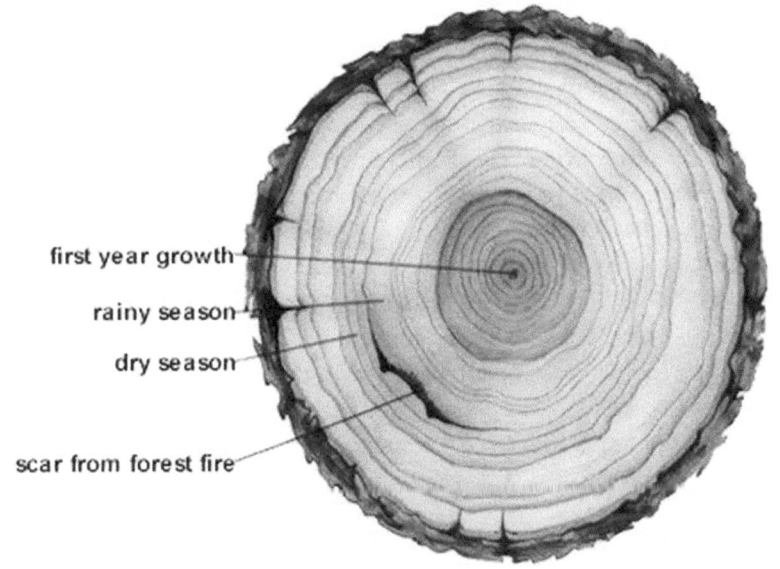

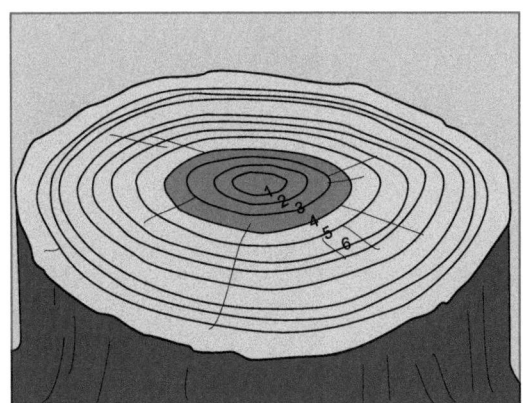

We can estimate the age of a tree by counting its rings.

Can you tell the age of the trees below?

© HERIOT-WATT UNIVERSITY

Q2:
Tree A =
....

Q3:
Tree B =

5.6 Learning points

Summary

- Plant organs are roots, stems and leaves.
- Leaf structure diagram show the upper epidermis, palisade mesophyll, spongy mesophyll, vein (consisting of xylem and phloem), lower epidermis, guard cells and stomata.
- The part of the plant involved in water transport is the xylem vessel.
- Water and minerals enter the plant through the root hairs and are transported in these dead xylem vessels.
- Xylem vessels are lignified to withstand the pressure changes as water moves through the plant.
- Transpiration is the process of moving water through the plant and its evaporation through the stomata.
- The rate of transpiration is affected by wind speed, humidity, temperature and surface area.
- Sugar is transported up and down the plant in living phloem.
- Phloem cells have sieve plates and associated companion cells.

5.7 End of topic test

End of topic test: Transport systems of plants Go online

Q4: Which of the following shows the passage of water through the tissues when it enters a plant?

a) root hair > xylem > spongy mesophyll
b) root hair > spongy mesophyll > xylem
c) spongy mesophyll > xylem > root hair
d) xylem > spongy mesophyll > root hair

....

The rate of transpiration in plants can be measured using the apparatus shown below. As the plant transpires, coloured water is drawn up the glass tube and its volume measured, over a set period of time, to give the rate of transpiration.

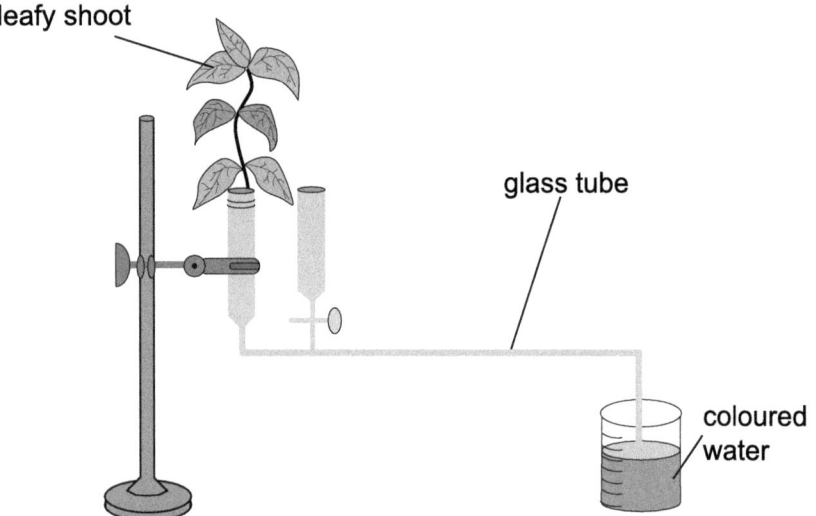

Changes in the surrounding environment can have an effect on the rate of transpiration. For each of the environmental changes listed below, state the effect this change has on the rate of transpiration.

Q5: Increase in humidity

....

Q6: Increase in temperature

....

Q7: Increase in wind speed

The graph below shows transpiration rates of two plants, P and Q.

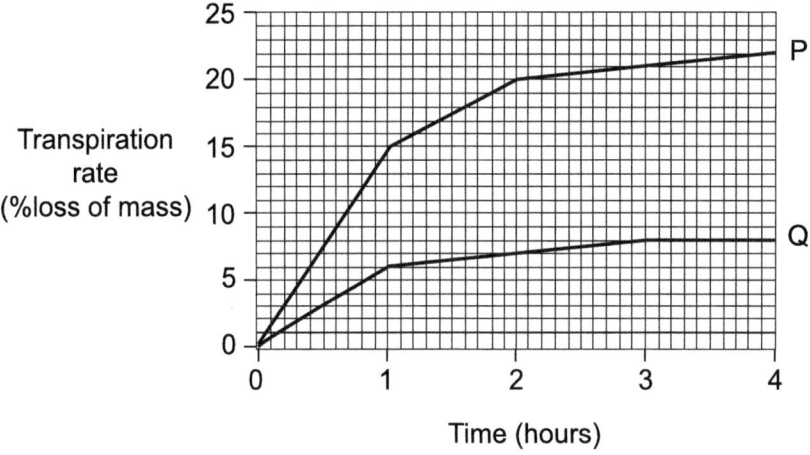

Q8: With reference to the number of stomata, suggest a reason for the different transpiration rates of plants for both P and Q.

- P has a greater number of stomata
- Q has fewer stomata
- Q has a greater number of stomata
- P has fewer stomata

....

Q9: Name the type of cells which control the opening and closing of stomata.

Q10:
The diagram below shows three parts of a plant.

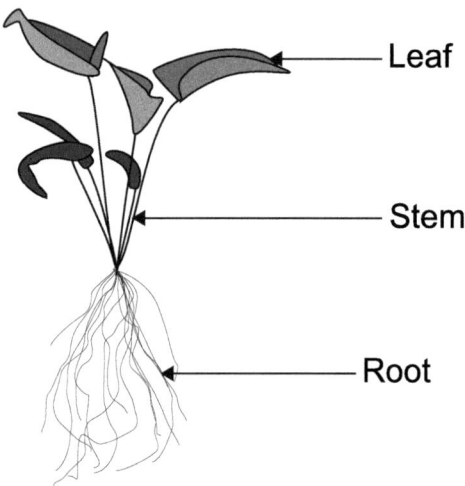

Describe the structures and processes involved as water moves through the plant from the soil to the air.

Water is _____ by root hairs by _____. Water travels _____ in the _____. Water travels to the _____ and _____.

Unit 2 Topic 6

Transport systems of animals

Contents

6.1 Blood cells 155
    6.1.1 Extension: Forming a scab 157
6.2 Immune system 157
6.3 Pathway of blood through the body 160
6.4 Structure and function of the heart 162
6.5 Blood vessels 166
6.6 Learning points 169
6.7 End of topic test 169

154 UNIT 2. MULTICELLULAR ORGANISMS

Learning objective

At the end of this topic you should be able to:

- State that blood contains red blood cells, white blood cells and plasma.
- State that blood transports oxygen, carbon dioxide and nutrients.
- Describe the structure of red blood cells as having no nucleus, biconcave shape, and containing haemoglobin.
- Explain that the specialisation of red blood cells aids them to transport oxygen in the form of oxyhaemoglobin.
- State that white blood cells are part of the immune system and are involved in destroying pathogens.
- Describe the two main types of cells involved (Phagocytes carry out phagocytosis and some lymphocytes produce antibodies to destroy pathogens)
- State that each antibody is specific to a particular pathogen.
- Describe the pathway of oxygenated and deoxygenated blood through heart, lungs and body.
- Label a diagram of heart with a right and left atria, ventricles, location of four valves as well as the aorta, vena cava, pulmonary artery, pulmonary vein and coronary arteries.
- State the function of each of these parts.
- Describe the structure and function of the blood vessels to be:
    - Arteries have thick, muscular walls, a narrow central channel and carry blood under high pressure away from the heart.
    - Veins have thinner walls, a wider channel and carry blood under low pressure back towards the heart. Veins contain valves to prevent backflow of blood.
    - Capillaries are thin walled and have a large surface area, forming networks at tissues and organs to allow efficient exchange of materials.

TOPIC 6. TRANSPORT SYSTEMS OF ANIMALS

6.1 Blood cells

The circulatory system consists of the heart (muscular pump) and a system of tubes called blood vessels. Together they carry blood to all parts of the body. The blood that you see when you have bloods taken at a hospital, or maybe even cut yourself, is a fluid made of different components.

1. Plasma:
   Plasma is a watery fluid that looks yellowish in colour. Plasma has substances like glucose, carbon dioxide and urea dissolved in it so that they can be transported around the body.

2. Platelets(*):
   Platelets are responsible for forming clots when a blood vessel is damaged, such as when we cut ourselves.

3. Red blood cells:
   The blood cells that transport oxygen. Red cells contain **haemoglobin** and it is this haemoglobin which allows them to associate with oxygen.

4. White blood cells:
   White blood cells are an important part of your immune system. They help fight infections by attacking bacteria, viruses, and germs.

(*The notes on platelets below is for information only/Extension work- Platelets are no longer part of the N5 course.)

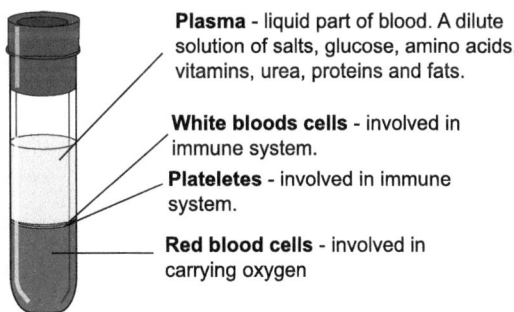

Plasma - liquid part of blood. A dilute solution of salts, glucose, amino acids, vitamins, urea, proteins and fats.

White bloods cells - involved in immune system.

Plateletes - involved in immune system.

Red blood cells - involved in carrying oxygen

A blood sample put through a centrifuge so that its composition can be seen more easily.

Composition of blood

Go online

Q1: Add the term from the list into the right line:

Blood component	Function
	Transports carbon dioxide, digested food, urea and hormones.
	Transports oxygen.
	Ingests pathogens and produces antibodies.
	Involved in blood clotting.

Word list: Red blood cells, Platelets, Plasma, White blood cells

The scanning EM microscope picture below has been colourised to highlight the different cell types. The red blood cell is in red, the platelet is in yellow and the lymphocyte is in blue.

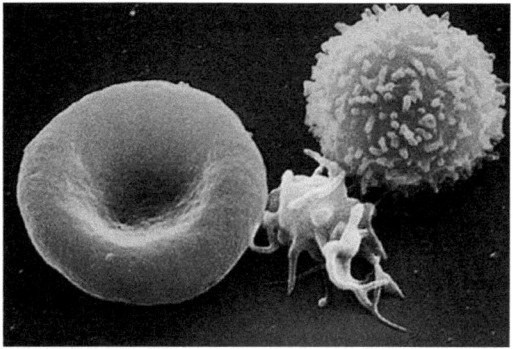

Video: EM images of blood composition

Go online

Watch this video showing EM images of blood composition.

https://youtu.be/9va0KPrVExs

Red blood cells move oxygen through the body within blood vessels for aerobic respiration to occur is body cells. It is essential that they are able to absorb oxygen in the lungs, move through different sized blood vessels, and release their oxygen to body cells. Red blood cells have adaptations that make them efficient at this:

- Haemoglobin — they contain this red protein as it combines with oxygen.

$$\text{Haemoglobin + Oxygen} \longrightarrow \text{Oxyhaemoglobin}$$

- No nucleus to contain more haemoglobin and maximise oxygen transportation.
- Small and flexible so enhance their mobility and fit through narrow blood **capillaries**.

- Biconcave shape (flattened disc shape) to increase their surface area for maximum oxygen absorption.

6.1.1 Extension: Forming a scab

Blood clotting:

If the skin becomes cut, the wounded area must be closed to minimise the loss of blood and the entry of pathogens into the open wound area. A scab will form to do this. Blood contains tiny platelets which are involved in blood clotting and scab formation.

Forming a scab:

When skin is wounded, platelets can release chemicals that cause specialised proteins to form an insoluble layer of fibres across the open wound. Platelets can also stick together to form lumps that get stuck in the fibrin mesh. Red blood cells also get stuck in the fibrin mesh, which can result in a clot and develops into a scab to protect the wound as it heals.

Blood cells: Questions Go online

Q2: Name the component of red blood cells that helps them to transport oxygen around the body.

....

Q3: Complete the word equation below

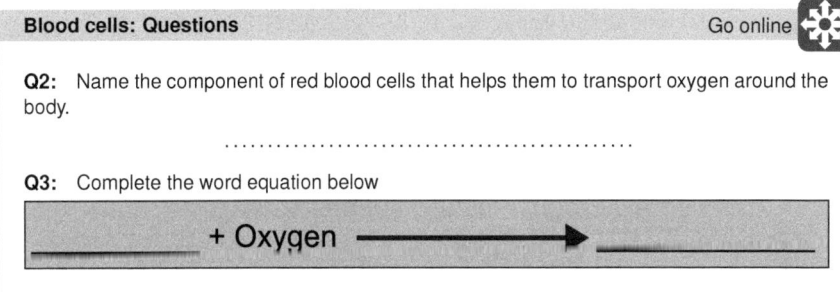

....

Q4: How do the following specialised features of red blood cells help their function?

- a) They contain haemoglobin.
- b) They have no nucleus.
- c) They are small and flexible.
- d) They have a biconcave shape.

List: so they can contain more haemoglobin, so that they can fit through narrow blood vessels, a red protein that combines with oxygen, to maximise their surface area for oxygen absorption

---

6.2 Immune system

The body's first line of defence means preventing any pathogens from entering in the first place. If a pathogen manages to get into the body, the second line of defence takes over. White blood cells are part of the immune system and are involved in destroying pathogens. The white blood cells do this in a variety of ways.

- ingest pathogens and destroy them

- produce antibodies to destroy pathogens
- produce antitoxins that neutralise the toxins released by pathogens.

There are two main types of cells involved.

Phagocytes	Engulfing pathogens in a process called phagocytosis
Lymphocytes	May produce antibodies specific to certain pathogens to destroy them.

Phagocytes can easily pass through blood vessel walls into the surrounding tissue and move towards pathogens or toxins. They then either:

- Ingest and absorb the pathogens or toxins.
- Release an enzyme to destroy them.
- Having absorbed a pathogen, the phagocytes may also send out chemical messages that help nearby lymphocytes to identify the type of antibody needed to neutralise them.

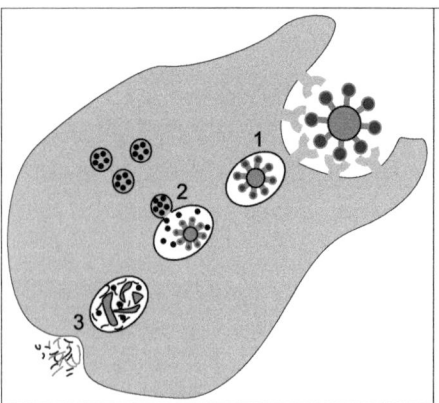

1. The invading pathogen releases chemicals which act as attractants for a white blood cell causing it to move towards the pathogen. The white blood cell then attaches to the surface of the pathogen and engulfs it to form a vesicle.

2. Lysosomes then move towards the vesicle and fuse with it. Enzymes within the lysosomes break down the pathogen and destroy it.

3. Waste products from the pathogen are absorbed into the cytoplasm.

Microscopic images of phagocytes Go online

Immune Cells Eating Bacteria (Phagocytosis)

https://www.youtube.com/watch?v=iZYLelJwe4w

Pathogens contain certain chemicals that are foreign to the body and are called antigens. Each lymphocyte carries a specific type of antibody — a protein that has a chemical 'fit' to a certain antigen (as shown in the diagram below).

TOPIC 6. TRANSPORT SYSTEMS OF ANIMALS

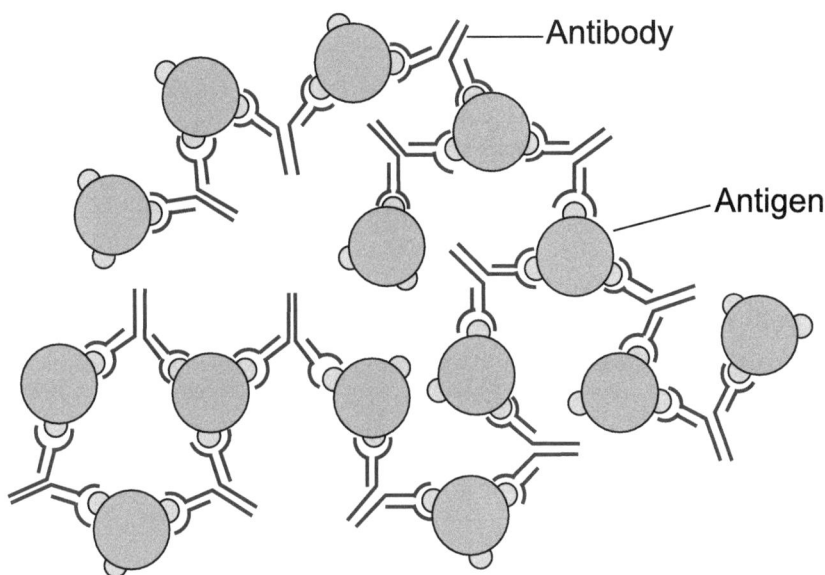

When a lymphocyte with the appropriate antibody meets the antigen, the lymphocyte reproduces quickly, and makes many copies of the antibody that neutralises the pathogen.

Antibodies neutralise pathogens in a number of ways:

- they bind to pathogens and damage or destroy them
- they coat pathogens, clumping them together so that they are easily ingested by phagocytes
- they bind to the pathogens and release chemical signals to attract more phagocytes

Lymphocytes may also release antitoxins that stick to the appropriate toxin and stop it damaging the body.

Microscopic images of lymphocytes Go online

Carry out a search for microscopic images of lymphocytes to view examples of these: https://bit.ly/2pejGa0

Immune system: Questions Go online

Q5: What do you call a micro-organism that causes disease?

....

Q6: What main process do phagocytes carry out?

....

© HERIOT-WATT UNIVERSITY

Q7: State the relationship between antibodies and pathogens.
- Each antibody is specific to a particular pathogen.
- Each antibody is specific to a particular phagocyte.
- Antibodies are not specific to particular pathogens.
- Antibodies are not specific to particular phagocytes.

6.3 Pathway of blood through the body

The circulatory system provides a pathway for oxygenated and deoxygenated blood to flow through the heart, lungs and body. To ensure all cells have a good blood supply, the network of blood vessels extend throughout the body.

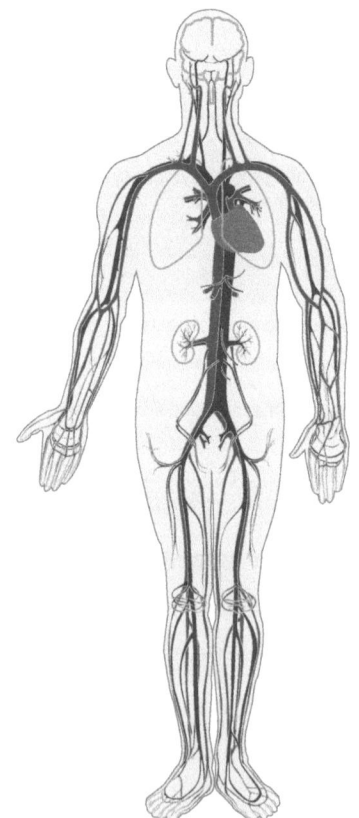

The heart pumps blood to the lungs so that oxygen can enter the blood. This oxygen rich blood is said to be 'oxygenated' and will then be pumped back to the heart so that it can be pumped out to the rest of the body. Oxygenated blood is shown in the diagram by the RED colour.

The BLUE colour in this diagram represents deoxygenated blood. This is blood that lacks oxygen and is being returned to the heart from the body.

TOPIC 6. TRANSPORT SYSTEMS OF ANIMALS

The diagram below shows the same transport route but with more detail. Now we can see that the left-hand side of the heart and body transports oxygenated blood (red coloured areas) and the right hand side transports deoxygenated blood (blue areas).

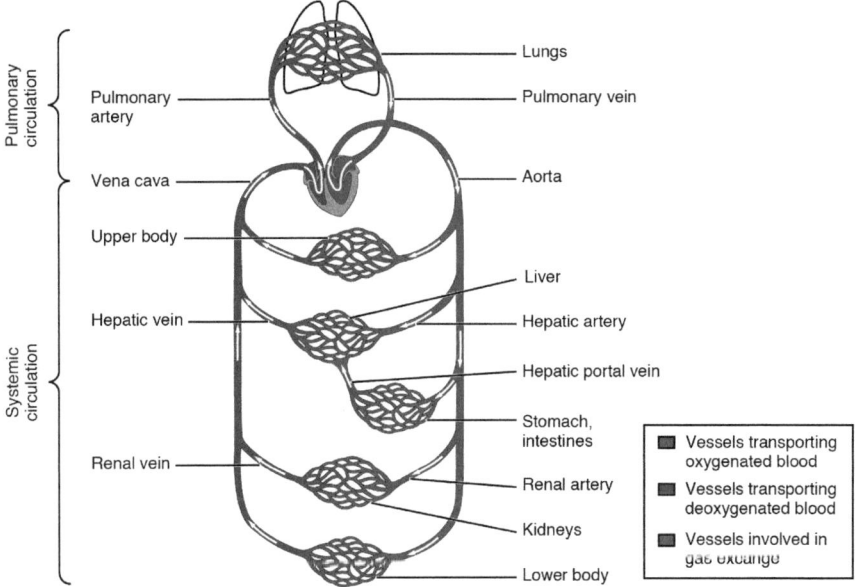

Arteries carry blood away from the heart towards an organ, while **veins** carry blood from an organ towards the heart.

The table lists the arteries and veins that are associated with the lungs, liver and kidneys as examples.

Organ	Towards the organ	Away from the organ
Lung	Pulmonary **artery**	Pulmonary vein
Liver	Hepatic artery	Hepatic vein
Kidney	Renal artery	Renal vein

© HERIOT-WATT UNIVERSITY

Pathways of blood through the body: Questions

Go online

Identify the correct word to complete the sentences below.

Q8:

Oxygen rich blood is said to be (oxygenated/ deoxygenated) and will then be pumped back to the heart so that it can be pumped out to the rest of the body. (Oxygenated/ Deoxygenated) blood is blood that lacks oxygen and is being returned to the heart from the body.

(Arteries/ Veins) carry blood away from the heart towards an organ, while (Arteries/ veins) carry blood from an organ towards the heart.

6.4 Structure and function of the heart

The heart is a muscle that is usually described, when linked to the circulatory system, as a double pump circulatory organ.

- The pulmonary circuit is the route between the heart and lungs.
- The systemic circuit is the route between the heart and the other organs.

The human heart has four chambers, two of which collect blood and the other two to pump blood. The upper chambers of the heart are called **atria** (or atrium when referring to a single chamber) and the lower chambers are called ventricles.

Atria collect blood from the blood vessels and pass the blood to the ventricles. The ventricles are the pumping chambers.

The diagram of heart below shows a very detailed labelled structure. At National 5 level, you should know the following:

TOPIC 6. TRANSPORT SYSTEMS OF ANIMALS

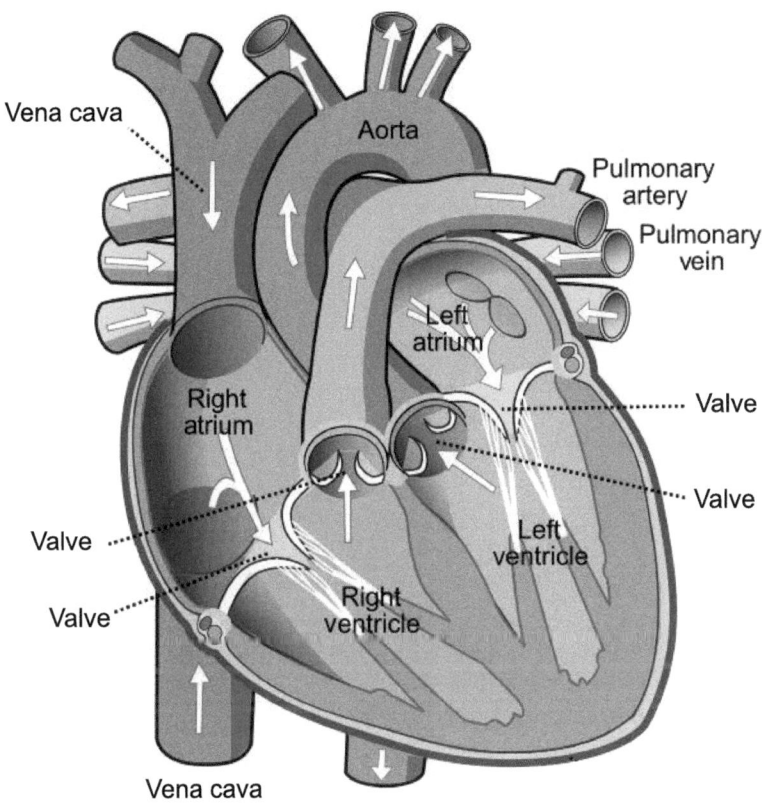

It is important that blood flows through the heart in one direction only and the valves of the heart ensure this occurs effectively. Valves can be found at different points:

1. Between the atria and ventricles
2. Where the pulmonary artery leaves the heart
3. Where the aorta leaves the heart.

Valves prevent the back flow of blood.

The coronary artery cannot been seen in the diagram above of the inside of the heart. This artery supplies the muscular wall of the heart itself with a rich supply of oxygenated blood.

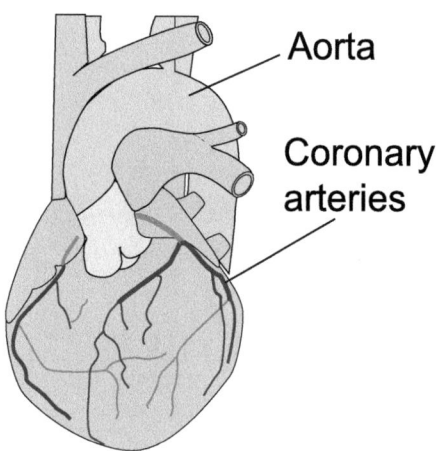

Aorta

Coronary arteries

Structures of the heart and their role:

Structure	Description/function
Aorta	The main artery which carries blood away from the heart
Atria	Upper chambers of the heart that pass blood to the lower ventricles
Pulmonary artery	Artery that carries deoxygenated blood from the heart to the lungs
Pulmonary vein	Vein that carries oxygenated blood from the lungs to the heart
Vena cava	Blood vessels that carry deoxygenated blood to the heart from the body
Ventricles	Lower chambers of the heart that receive blood from the upper atria

The pathway that blood takes to flow through the heart and body is:

TOPIC 6. TRANSPORT SYSTEMS OF ANIMALS

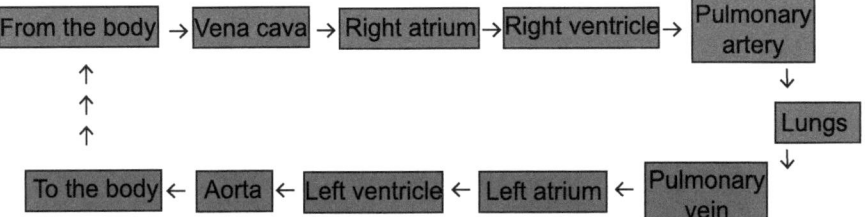

Key point

Remember: the blue areas indicate the pathway of deoxygenated blood and the red areas are the pathway for oxygenated blood.

Heart beating normally sound effect   Go online

When you listen to your heart with a stethoscope, the sound that you hear is the heart valves closing after blood passes through them.

https://youtu.be/gJpT_wHZeF8

Video: The working of the human heart   Go online

Watch this video about the human circulatory system.

https://youtu.be/_qmNCJxpsr0

(Please remember that you are not required to know the names of the valves.)

Video: Heart dissection   Go online

For those of you who feel uncomfortable watching a real heart dissection, here is a virtual heart dissection (computer animated version).

https://youtu.be/9l-XcW0XXzg

Structure of the heart: Questions   Go online

Q9:
Match the following terms to their correct definition.

Term	Meaning
Aorta	Artery that carries deoxygenated blood from the heart to the lungs
Atria	Blood vessels that carry deoxygenated blood to the heart from the body
Pulmonary artery	Lower chambers of the heart that receive blood from the upper artria
Pulmonary vein	The main artery which carries blood away from the heart
Vena cava	Upper chambers of the heart that pass blood to the lower ventricles
Ventricles	Vein that carries oxygenated blood from the lungs to the heart

....

Q10: Name the four chambers of the heart.

....

Q11: What is the name of the main artery which carries blood away from the heart?

....

Q12: Name the blood vessels that carry deoxygenated blood to the heart from the body.

6.5 Blood vessels

Arteries have thick, muscular walls, a narrow central channel and carry blood under high pressure away from the heart.

Veins have thinner walls, a wider channel and carry blood under low pressure back towards the heart. Veins contain valves to prevent backflow of blood.

Capillaries are thin walled and have a large surface area, forming networks at tissues and organs to allow efficient exchange of materials.

TOPIC 6. TRANSPORT SYSTEMS OF ANIMALS

Veins	Capillaries	Arteries
• Carry blood to the heart (always deoxygenated apart from the pulmonary vein which goes from the lungs to the heart). • Have thin walls. • Have larger passage ways for blood (internal lumen). • Contain blood under low pressure. • Have valves to prevent blood flowing backwards.	• Found in the muscles and lungs. • Microscopic — one cell thick • Very low blood pressure. • Where gas exchange takes place — oxygen passes through the capillary wall and into the tissues, while carbon dioxide passes from the tissues into the blood.	• Carry blood away from the heart (always oxygenated with the exception of the pulmonary artery which transports blood from the heart to the lungs). • Have thick muscular walls. • Have small central spaces for blood (internal lumen). • Contain blood under high pressure.

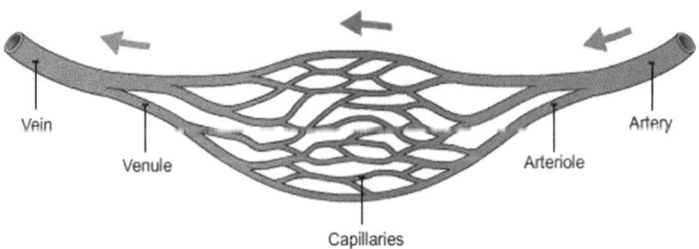

A different view, showing the insides of each vessel can be seen in the diagram below.

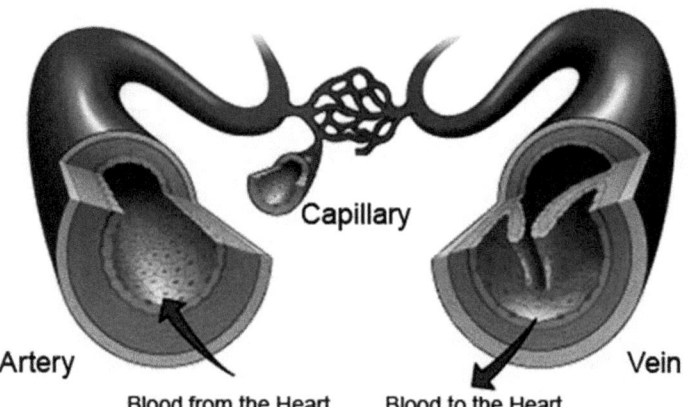

Artery — Blood from the Heart
Capillary
Vein — Blood to the Heart

Blood vessels: Questions Go online

Complete the questions below to identify which of them gives a description of arteries, capillaries and veins. Type the name of the vessel into the box.

Q13: Name the blood vessels which:

- Carry blood to the heart.
- Have thin walls.
- Contain blood under low pressure.
- Have valves to prevent blood flowing backwards.

....

Q14: Name the blood vessels which:

- Are found in the muscles and lungs.
- Are microscopic — one cell thick.
- Have very low blood pressure.
- Where gas exchange takes place.

....

Q15: Name the blood vessels which:

- Carry blood away from the heart.
- Have thick muscular walls.
- Contain blood under high pressure.

6.6 Learning points

Summary

- Blood contains red blood cells, white blood cells and plasma and transports oxygen, carbon dioxide and nutrients.
- Red blood cells have no nucleus, biconcave shape, and containing haemoglobin.
- Explain that the specialisation of red blood cells aids them to transport oxygen in the form of oxyhaemoglobin.
- White blood cells are part of the immune system and are involved in destroying pathogens.
- Phagocytes and lymphocytes are white blood cells. Phagocytes carry out phagocytosis and some lymphocytes produce antibodies to destroy pathogens.
- Pathway of oxygenated and deoxygenated blood through heart, lungs and body.
- A diagram of heart should show the right and left atria, ventricles, location of four valves, location of associated blood vessels (aorta, vena cava, pulmonary artery, pulmonary vein and coronary arteries).
- Arteries have thick, muscular walls, a narrow central channel and carry blood under high pressure away from the heart.
- Veins have thinner walls, a wider channel and carry blood under low pressure back towards the heart. Veins contain valves to prevent backflow of blood.
- Capillaries are thin walled and have a large surface area, forming networks at tissues and organs to allow efficient exchange of materials.

6.7 End of topic test

End of topic test: Transport systems of animals Go online

The table below gives information about features of three different types of blood vessel:

Q16:
Complete the table by writing the name of the missing types of blood vessels in the empty boxes.

Name of blood vessel	Diameter of central channel (mm)	Thickness of wall (mm)
	25.0	1.25
Capillary	0.008	0.001
	20.0	2.2

....

Q17: Of all the blood vessels, blood capillaries are best adapted for exchanging gases. Using the information in the table, give a reason for this.

The diagram below shows the heart and associated blood vessels.

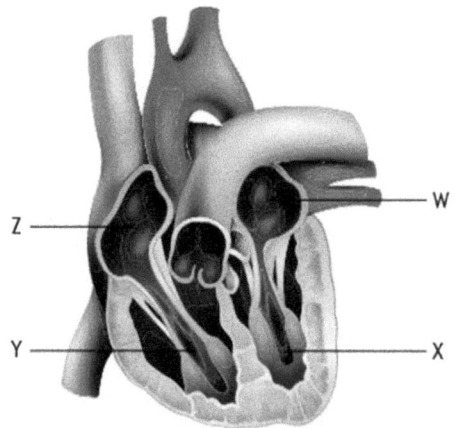

Q18: Which of the following statements is correct?

a) W is the left atrium which receives blood from the body.
b) X is the left ventricle which pumps blood to the body.
c) Y is the right atrium which receives blood from the lungs.
d) Z is the right ventricle which pumps blood to the lungs.

....

Q19: Name the blood vessel which supplies the heart muscle with blood.

The following sequence shows part of the blood flow through the body.

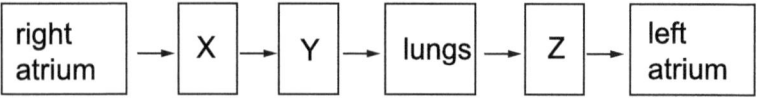

Q20: Which line in the table below identifies X, Y and Z?

	X	Y	Z
A	right ventricle	pulmonary vein	pulmonary artery
B	right ventrice	pulmonary artery	pulmonary vein
C	pulmonary vein	pulmonary artery	right ventricle
D	pulmonary article	right ventricle	pulmonary vein

....

Complete the pathway by typing the names into the box to show the correct pathway of blood.

Q21:
Wordlist: left ventricle, vena cava, pulmonary artery, right ventricle, pulmonary vein

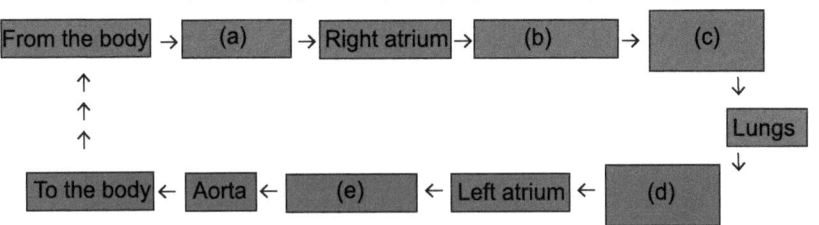

Unit 2 Topic 7

Absorption of materials

Contents

7.1 The need for transport 174
7.2 The role of capillary networks 175
7.3 The role of absorption surfaces in the body 176
7.4 Gas exchange in the lungs 177
7.5 Absorption in the small intestine 180
7.6 Learning points 184
7.7 End of topic test 184

Learning objective

At the end of this topic you should be able to:

- State that oxygen and nutrients from food must be absorbed into the bloodstream to be delivered to cells for respiration. Waste materials, such as carbon dioxide, must be removed from cells into the bloodstream.

- State that tissues contain capillary networks to allow the exchange of materials at cellular level.

- Describe the surfaces involved in the absorption of materials (i.e. they have certain features in common: large surface area, thin walls, extensive blood supply).

- State that the surfaces involved in the absorption of materials increases the efficiency of absorption.

- State that lungs are gas exchange organs.

- Describe the structure of lungs (they consist of a large number of alveoli providing a large surface area).

- State that oxygen and carbon dioxide are absorbed through the thin alveolar walls.

- State that nutrients from food are absorbed into the villi in the small intestine.

- Describe the structure of a villi (the large number of thin walled villi provides a large surface area. Each villus contains a network of capillaries to absorb glucose and amino acids and a lacteal to absorb fatty acids and glycerol).

7.1 The need for transport

It is essential that oxygen and nutrients from food are moved around the body in the bloodstream. They must be absorbed into the bloodstream and then effectively transported into the cells of the body. It is also vital that any waste materials, such as carbon dioxide and urea, are absorbed from body cells, transported away by the bloodstream and removed from the body efficiently. Below is a diagram to summarise the various transport functions of the blood.

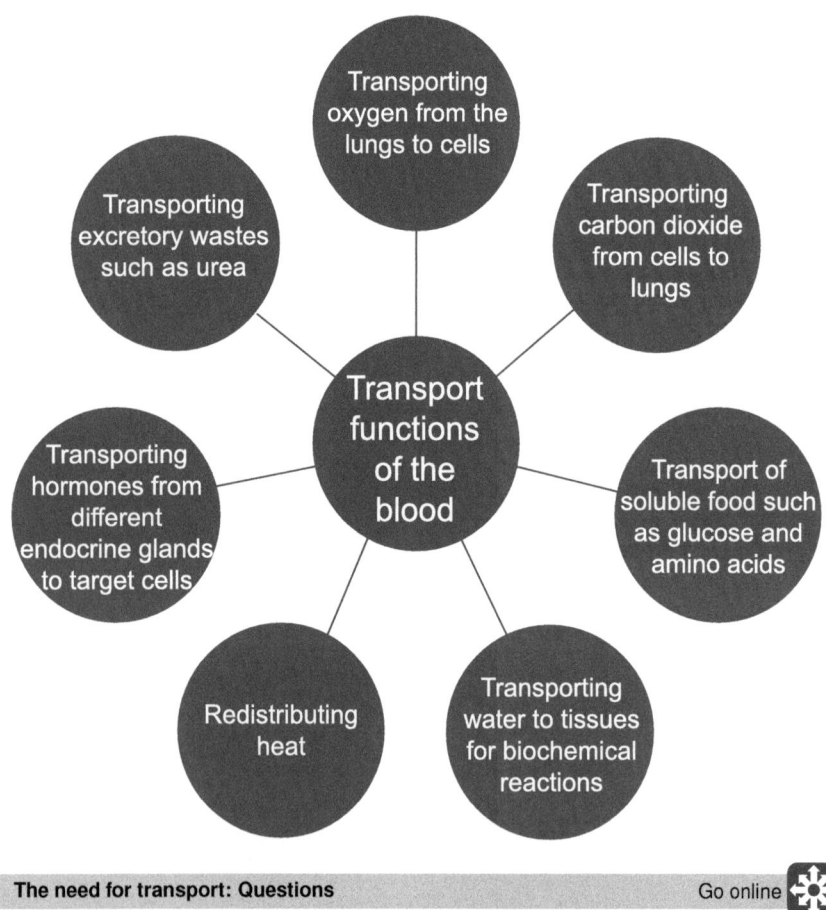

The need for transport: Questions   Go online

Give an example of what the blood could transport in the following cases:

Q1: Soluble food

....

TOPIC 7. ABSORPTION OF MATERIALS

Q2: Excretory wastes
....

Q3: To the lungs from body cells
....

Q4: From the lungs to body cells

7.2 The role of capillary networks

Capillaries are the blood vessels that have the appropriate structure to carry out the function of exchanging materials between the blood and the cells of the body. Capillaries are the blood vessels that are most numerous in the body as well as the only vessels that come into close contact with tissues and organs all over the body. The structures of capillaries make them ideal to efficiently exchange materials, especially when they form a capillary network.

Feature	Purpose
Large surface area	More materials can pass by diffusion into cells.
Thin walls	Very quick diffusion of materials into cells.
Extensive blood supply	More opportunities for body cells to contact a capillary network.

Capillaries

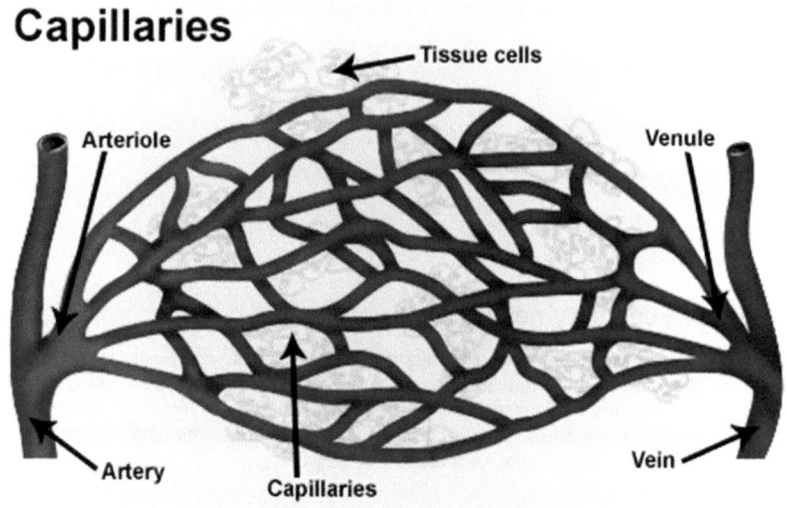

The role of capillary networks: Questions

Describe how the 3 features of a capillary network help its function.

Q5: Large surface area

....

Q6: Thin walls

....

Q7: Extensive blood supply

7.3 The role of absorption surfaces in the body

To be efficient at absorbing materials, absorption surfaces all over the body must have the same features as a capillary network;

1. Large surface area.
2. Thin walls.
3. A very good supply of blood.

In the lungs, the gases oxygen and carbon dioxide are exchanged between the blood and **alveoli** (air sacs).

In the small intestine, the products of digestion diffuse through its wall to enter the **lacteal** and the blood stream.

Any surface must be able to absorb the necessary products for vital cell processes such as respiration and many other biochemical reactions, as well as nourishing and repairing the body.

The role of absorption surfaces in the body: Question   Go online

Q8: Which of the following would allow more efficient gas exchange in the lungs?

a) low number of thin walled alveoli
b) high number of thin walled alveoli
c) low number of thick walled alveoli
d) high number of thick walled alveoli

7.4 Gas exchange in the lungs

The gas exchange organs in the body are the lungs which are part of the human respiratory system. The diagrams below show the internal structure of the lungs and the direction of the flow of air.

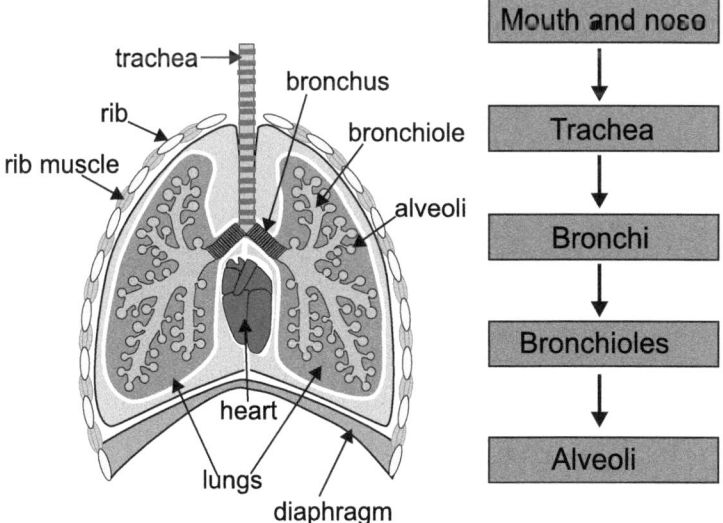

The mouth and nose contain air passage ways that are connected to the trachea. Each lung receives air from the trachea via two bronchi (singular is bronchus). Smaller air tubs called bronchioles are lead to millions of microscopic air sacs called alveoli.

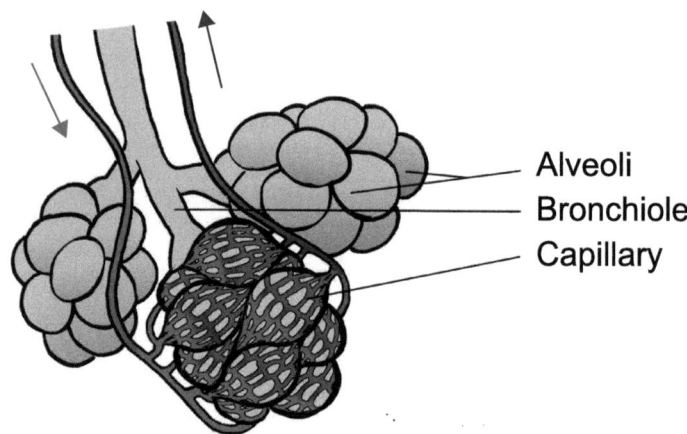

Alveoli
Bronchiole
Capillary

Alveoli provide a large surface area for gas exchange to take place. Carbon dioxide and oxygen diffuse across the thin walled alveoli easily and very rapidly. The alveoli have a dense network of blood capillaries around them to transport oxygen to the body cells and carbon dioxide from body cells to the lungs.

Gas	Movement/direction
Oxygen	From a high concentration in alveoli to a low concentration in blood capillaries
Carbon dioxide	From a high concentration in blood capillaries to a low concentration in alveoli

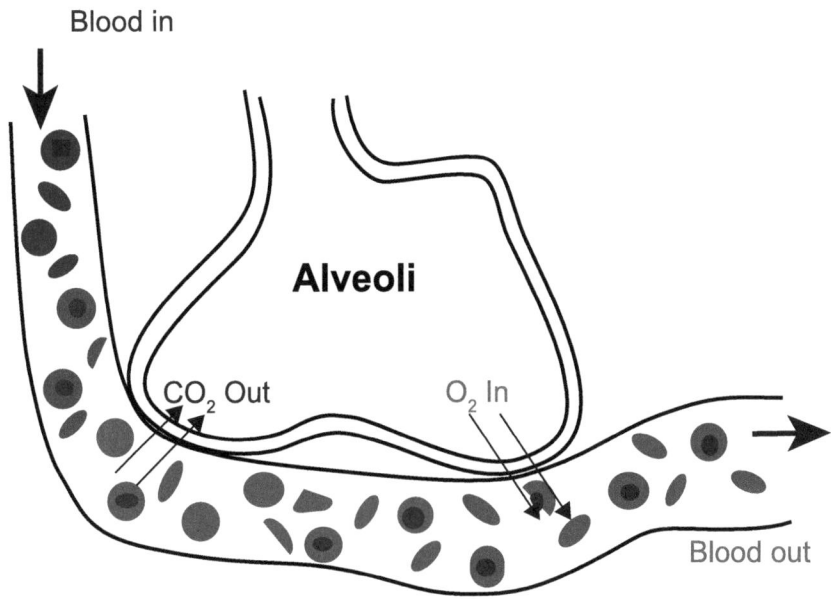

| Video: Alveoli Gas Exchange | Go online |

Watch this video about gas exchange in the lungs.

https://youtu.be/mZvzl8KH6iI

Gas exchange in the lungs: Questions

Go online

Q9: Use the diagram of the lungs below to identify the structures labelled A, B, C & D:

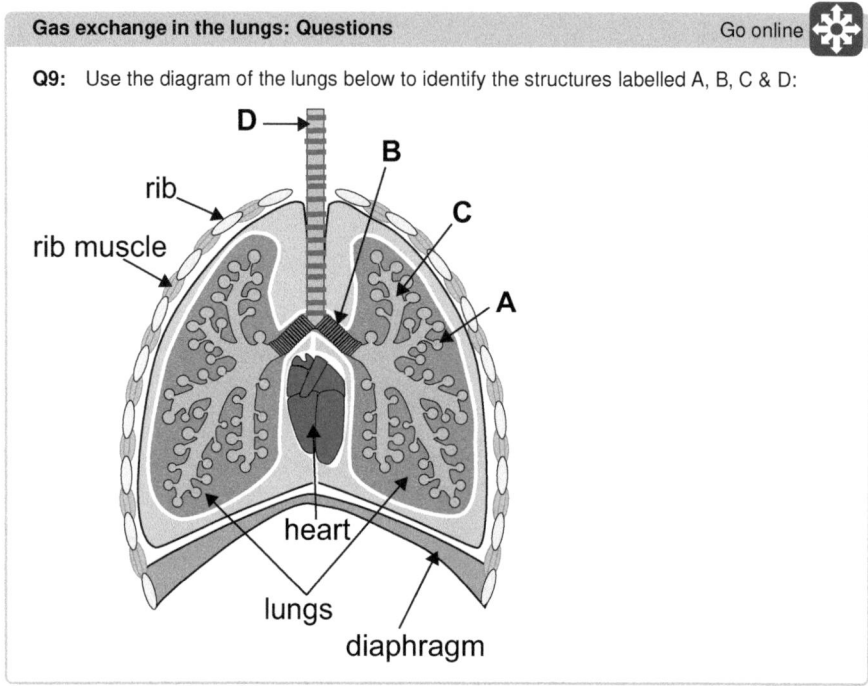

7.5 Absorption in the small intestine

Many of the food molecules we eat are insoluble in their original form and must be digested to be absorbed is a soluble form. Enzymes present throughout the digestive system aid the breakdown of large insoluble food molecules to small soluble molecules so that they can dissolve and diffuse across absorption surfaces.

Food travels through the alimentary canal by a process called **peristalsis**. Muscles from the mouth to the anus (the alimentary canal) contract behind the food to squeeze it, while the muscles in front of the food relax to allow it to pass from one end of the canal to the other. The food will come into contact with various enzymes in different parts of the digestive system until it reaches the small intestine.

TOPIC 7. ABSORPTION OF MATERIALS

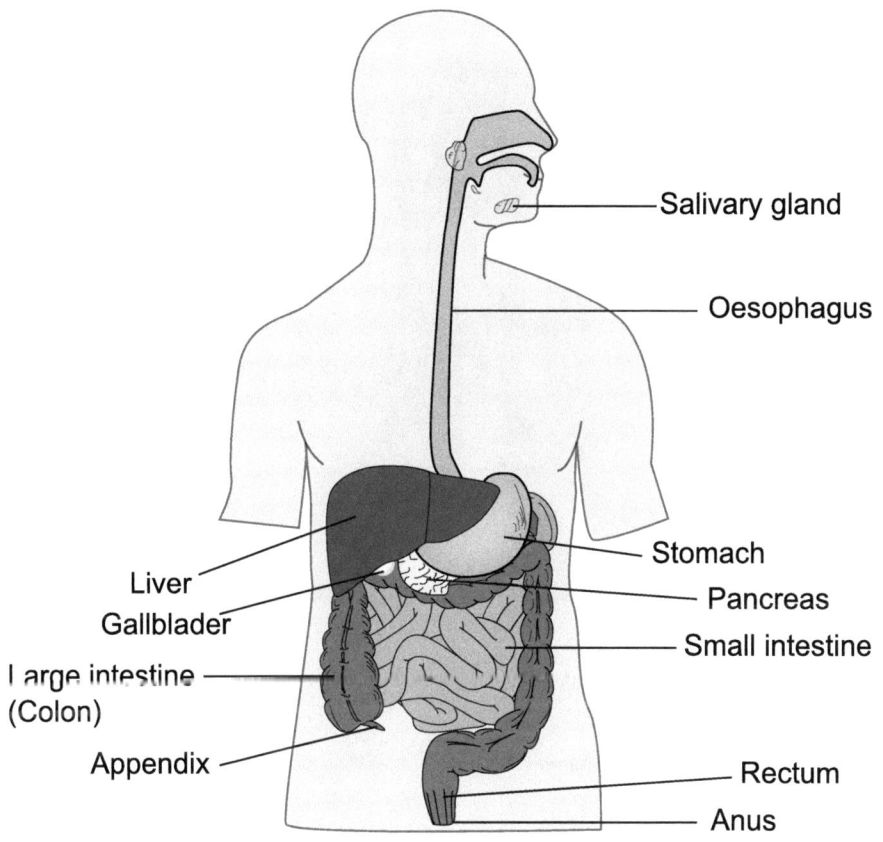

The role of the small intestine is to absorb the final products of digestion. The small intestine has a villus lining to allow it to do this efficiently.

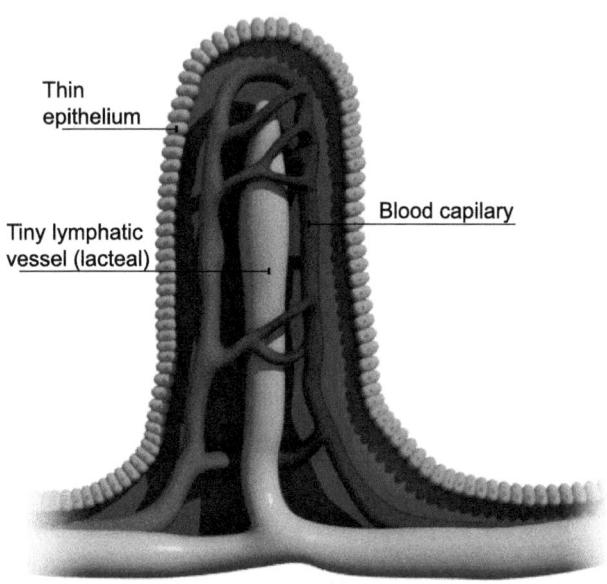

The long, inner lining of the small intestine has many folds with many **villi** looking like finger-like projections.

Micrographs of the small intestine	Go online

https://binged.it/2rpQRZ0

Each villi has features common to all absorption surfaces.

a) The thin wall is only one cell thick to allow any dissolved molecules to pass through rapidly by diffusion.
b) The capillary network is in close contact to provide a good blood supply and the transport of amino acids (from proteins) and glucose (from carbohydrates).

The only feature of the villi that is different is the lacteal. The products of fat digestion, fatty acids and glycerol, can not be absorbed into the blood and must take an alternative transport system. The central lacteal provides transport from the small intestine to the **lymphatic** system.

Absorption in the small intestine: Questions

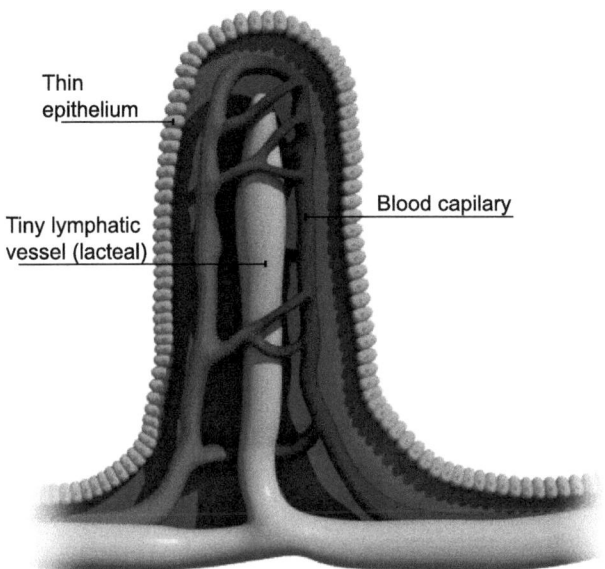

Q10: Which food molecules are absorbed into the capillaries of the villus?

a) Fatty acids and glycerol
b) Amino acids and glycerol
c) Amino acids and glucose
d) Fatty acids and glucose

....

Q11: Which food molecules are absorbed into the lacteal of the villus?

a) Fatty acids and glycerol
b) Amino acids and glycerol
c) Amino acids and glucose
d) Fatty acids and glucose

7.6 Learning points

Summary

- Oxygen and nutrients from food must be absorbed into the bloodstream to be delivered to cells for respiration.

- Waste materials, such as carbon dioxide, must be removed from cells into the bloodstream.

- Tissues contain capillary networks to allow the exchange of materials at cellular level.

- Surfaces involved in the absorption of materials have certain features in common to increase the efficiency of absorption:
    - large surface area
    - thin walls
    - extensive blood supply.

- Lungs are gas exchange organs.

- Lungs consist of a large number of alveoli providing a large surface area.

- Oxygen and carbon dioxide are absorbed through the thin alveolar walls to or from the many blood capillaries.

- Nutrients from food are absorbed into the villi in the small intestine.

- The large number of thin walled villi provides a large surface area.

- Each villus contains a network of capillaries to absorb glucose and amino acids and a lacteal to absorb fatty acids and glycerol.

7.7 End of topic test

End of topic test: Absorption of materials Go online

The diagram shows a site of gas exchange in the lungs.

Q12:

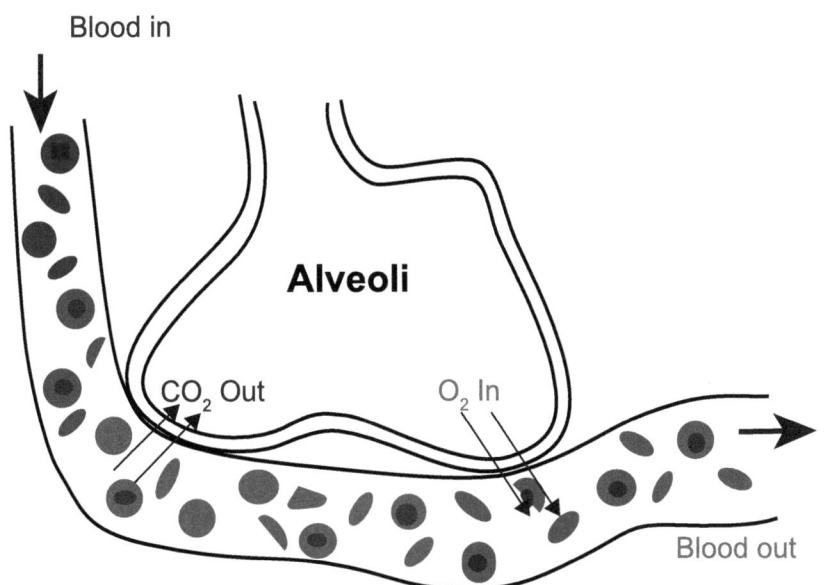

The table shows the relative concentration of oxygen and carbon dioxide in three cell types. Complete the table to show the missing information.

Cell type	Relative concentration of gases Oxygen	Relative concentration of gases Carbon dioxide
Red blood cell	Low	
Alveolus cell		Low
Capillary wall cell	Medium	Medium

Q13: Describe the pathway that oxygen would take when moving between these cell types.

Q14: Describe two features of the lungs which improve the efficiency of gas exchange.

Q15: Matching exercise:

Term	Meaning
Lacteal	Waves of muscular contractions that help food move through the alimentary canal.
Lymph	Finger-like projections in the small intestine that provide a large surface area for absorbing food.
Peristalsis	Vessel in the villi that is responsible for transporting fats.
Villi	Liquid that transports the products of fat digestion from the lacteal.

....

Q16: Choose the correct word:

Food travels through the alimentary canal by a process called (digestion/ peristalsis).

Muscles in the alimentary canal (contract/ relax) behind the food to squeeze it, while the muscles in front of the food (contract/ relax) to allow it to pass from one end of the canal to the other.

The food will come into contact with various enzymes in different parts of the digestive system until it reaches the (large/small) intestine.

Enzymes present throughout the digestive system aid the breakdown of large (soluble/ insoluble) food molecules to small (soluble/ insoluble) molecules so that they can dissolve and diffuse across absorption surfaces.

Unit 2 Topic 8
Multicellular organisms test

Multicellular organisms test

Go online

Producing new cells

Q1: Match the following terms to their correct definition.

Term	Meaning
Chromatid	A process of cell division that produces two genetically identical daughter cells.
Chromosome compliment	Protein threads that pull chromatids apart during mitosis.
Chromosome	The number of chromosomes found in a cell.
Diploid	Replicated copy of a chromosome.
Equator	Codes for all of an organisms characteristics.
Mitosis	Middle position of a cell where chromosomes align and attach to spindle fibres in mitosis. Composed of DNA.
Spindle fibres	A cell that contains a double set of chromosomes.

....

Q2: Choose the correct word to complete the sentences.

Stem cells are (unspecialised/specialised) cells found in (plants/animals) .

....

Q3: Choose two uses of stem cells after they divide.

- One use is to make cells that develop into generic cells.
- One use is to make more stem cells (self-renew).
- One use is to track damaged tissue.
- One use is to make cells that develop into specialised cells.

....

Q4: State the function of red blood cells and describe how they are specialised to suite this role.

TOPIC 8. MULTICELLULAR ORGANISMS TEST

Control and communication

Q5: Match the following terms to their correct definition.

Term	Meaning
Brain	Nerve cell that are found in the CNS where they connect with other neurons.
Central Nervous System (CNS)	Part of the nervous system made up of the brain and spinal chord.
Cerebellum	Section of the brain that controls memory, conscious thoughts, intelligence and emotions.
Cerebrum	Organ of the central nervous system that controls vital functions.
Inter neuron	Section of the brain that controls breathing and heart rate.
Medulla	Section of the brain that controls coordination, movements and balance.

....

Q6: Match the following terms from the word list to their correct definition:

Neurones	Function
	are nerve cells that carry electrical impulses from sense organs to CNS.
	are nerve cells that are found in the CNS where they connect with other neurons.
	are nerve cells that carry electrical impulses from the CNS to muscles and glands (effectors).

Wordlist: Motor, Sensory, Inter

Reproduction

Q7:
The diagram below shows the structure of a flower.

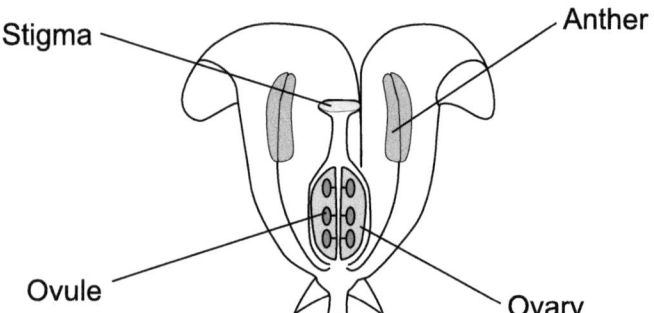

State the function of the labelled structures.

- Anther
- Pollen grains
- Ovary
- Ovules

....

Q8: Match the following stages of the process of plant fertilisation to their correct description.

STAGE	Description
Stage 1	The nucleus inside the pollen grain starts to make its way down the inside of the tube.
Stage 2	The tip of the pollen tube will then burst to release the male gamete so that it may fuse with the female gamete and fertilisation can take place.
Stage 3	The pollen grain begins to grow a pollen tube through the tissues of the style and towards the ovary.
Stage 4	When the end of the pollen tube reaches and ovule in the ovary, it enters via a tiny hole.

....

Q9: Choose the correct word to complete the sentences.

Body cells are (haploid/diploid) in multicellular organisms and are (haploid/diploid) in gametes (sex cells).

Diploid means that each cell contains (one/two) sets of matching chromosomes.

Haploid means that each cell contains (one/two) set of chromosomes.

TOPIC 8. MULTICELLULAR ORGANISMS TEST

Variation and inheritance

Q10:
Predict the genotypes and phenotypes of hamster fur type from a cross between a normal fur hamster (RR) and a rex fur hamster (rr)

Parental genotypes	r	r
R		
R		

....

Q11: State the difference between discrete and continuous variation.

....

Q12: Match the following terms to their correct definition.

Term	Meaning
Genotype	Type of inheritance involving several genes acting together.
Heterozygous	Two alleles the same for a genotype ie. AA or aa.
Homozygous	Two different alleles of a genotype ie. Aa or Bb.
Phenotype	The particular alleles that an organisms has for a genotype.
Polygenic	The form of a gene which will only be expressed if the genotype is homozygous.
Recessive	The physical appearance expressed by an organisms due to their genotype

Transport systems of plants

Q13:
The diagram below shows three parts of a plant.

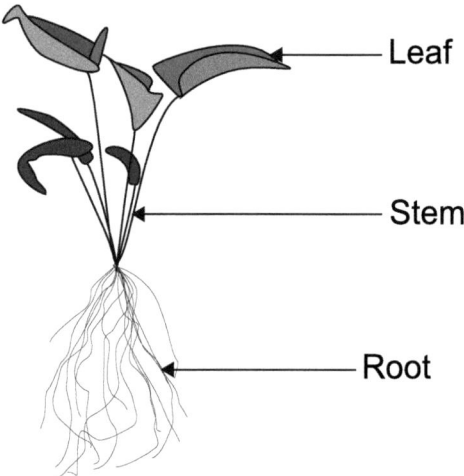

Leaf

Stem

Root

Name the processes involved as water moves through the plant from the soil to the air.

....

Q14: Describe the role of guard cells.

....

Q15: Match the following terms from the word list to their correct definition.

Cell type	Function
	Sugar is transported up and down the plant in this living tissue.
	Water and minerals enter the plant through the root hairs and are transported in these.

*Wordlist:*Xylem, Phloem

....

Q16:
Identify structures (a) and (b) in the passage below.

(a) vessels are lignified to withstand the pressure changes as water moves through the plant.
(b) have sieve plates and associated companion cells.

TOPIC 8. MULTICELLULAR ORGANISMS TEST

Transport systems of animals

Q17:
The diagram below shows the heart and associated blood vessels.

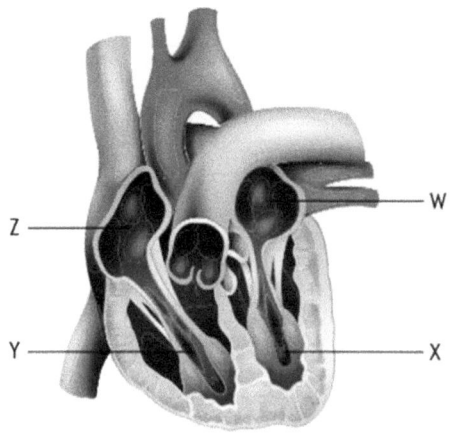

Name the four chambers

....

Q18: Match the following terms to their correct definition.

Blood component	Function
Plasma	Transports oxygen.
Red blood cells	Ingests pathogens and produces antibodies.
White blood cells	Transports carbon dioxide, digested food, urea and hormones.

Absorption of materials

Q19: Use the word list of the following features and match them to their purpose.

Worldlist: Large surface area, thin walls, extensive blood supply

Feature	Purpose
	Much more materials can pass by diffusion into cells.
	Very quick diffusion of materials into cells.
	More opportunities for body cells to contact a capillary network.

....

Q20: Match the following terms to their correct definition.

Term	Meaning
Alveoli	Vessel in the villi that is responsible for transporting fats.
Cartilage	Liquid that transports the products of fat digestion from the lacteal.
Lacteal	Finger-like projections in the small intestine that provide a large surface area for absorbing food.
Lymph	Tiny sacs for gas exchange in lungs.
Villi	Flexible tissue in the trachea to keep the airway open.

Life on Earth

1 Ecosystems		**197**
1.1	Definitions of ecological terms	198
1.2	Interactions of organisms in food webs	200
1.3	Niche	201
1.4	Competition	202
1.5	Learning points	205
1.6	End of topic test	206
2 Distribution of organisms		**211**
2.1	Abiotic and biotic factors	212
2.2	Measuring abiotic factors	213
2.3	Sampling of plants and animals	215
2.4	Paired-statement keys	217
2.5	Indicator species	219
2.6	Learning points	221
2.7	Extended response	222
2.8	End of topic test	223
3 Photosynthesis		**227**
3.1	Reactions of photosynthesis	228
3.2	Limiting factors	229
3.3	Learning points	232
3.4	Extended response	232
3.5	Extension materials	233
3.6	End of topic test	235
4 Energy in ecosystems		**239**
4.1	Energy in ecosystems	240
4.2	Learning points	242
4.3	End of topic test	243

5 Food production 245
5.1 The increasing human population 246
5.2 Fertilisers 246
5.3 Pesticides 248
5.4 Learning points 252
5.5 Extended response 253
5.6 Extension materials 253
5.7 End of topic test 255

6 Evolution of species 257
6.1 Mutation 258
6.2 Natural selection 259
6.3 Speciation 262
6.4 Learning points 265
6.5 Extended response 265
6.6 Extension materials 266
6.7 End of topic test 267

7 Life on Earth test 269

Unit 3 Topic 1

Ecosystems

Contents
1.1 Definitions of ecological terms 198
1.2 Interactions of organisms in food webs 200
1.3 Niche 201
1.4 Competition 202
1.5 Learning points 205
1.6 End of topic test 206

Learning objective

By the end of this topic you should be able to:

- define the terms: species, biodiversity, population, producer, consumer, herbivore, carnivore, omnivore, predator, prey, food chain, food web and ecosystem;

- describe the interactions of organisms in food webs;

- describe the effects of removal of organism(s) from a food web;

- describe an organism's niche;

- state that competition occurs when resources are in short supply;

- describe the difference between interspecific competition and intraspecific competition;

- state that intraspecific competition is more intense than interspecific competition.

1.1 Definitions of ecological terms

Ecology is a branch of biology which relates to relationships between organisms and their habitat. This area of science uses many biological terms.

A species is a group of organisms which can interbreed and produce **fertile** offspring. A labrador and a poodle are members of the same species since the offspring they produce (labradoodles) are fertile. A horse and a donkey are different species since the offspring they produce (mules) are not fertile. The term population describes all the members of one species within an ecosystem. Biodiversity refers to the range of species within an ecosystem. Ecosystems with high biodiversity have many different species and ecosystems with low biodiversity have few species. An ecosystem consists of all the organisms (the community) living in a particular habitat and the non-living components with which the organisms interact.

Organisms can be classed according to their feeding habits:

- producers are organisms which produce their own food whereas consumers are organisms which consume other organisms for food;
- herbivores only consume plant matter, carnivores only consume animal matter and omnivores consume both plant and animal matter;
- predators are animals which hunt other animals for food and prey are animals which are hunted and killed by predators.

A food chain is a diagram which shows a simple feeding relationship between organisms.

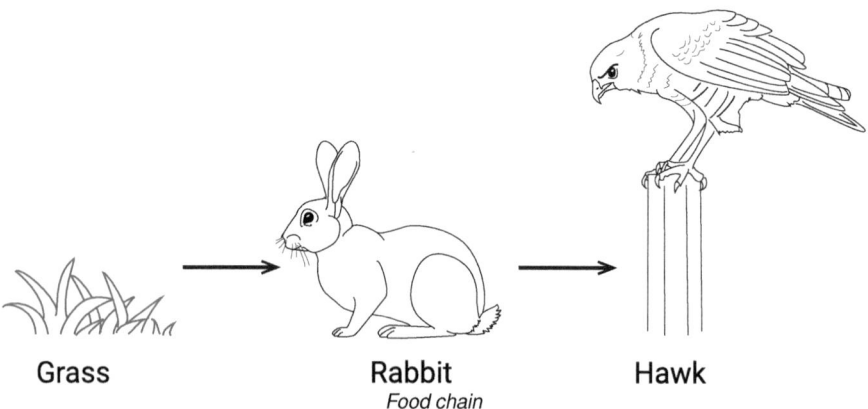

Grass → Rabbit → Hawk

Food chain

TOPIC 1. ECOSYSTEMS

A food web is a diagram which shows the interconnection of food chains.

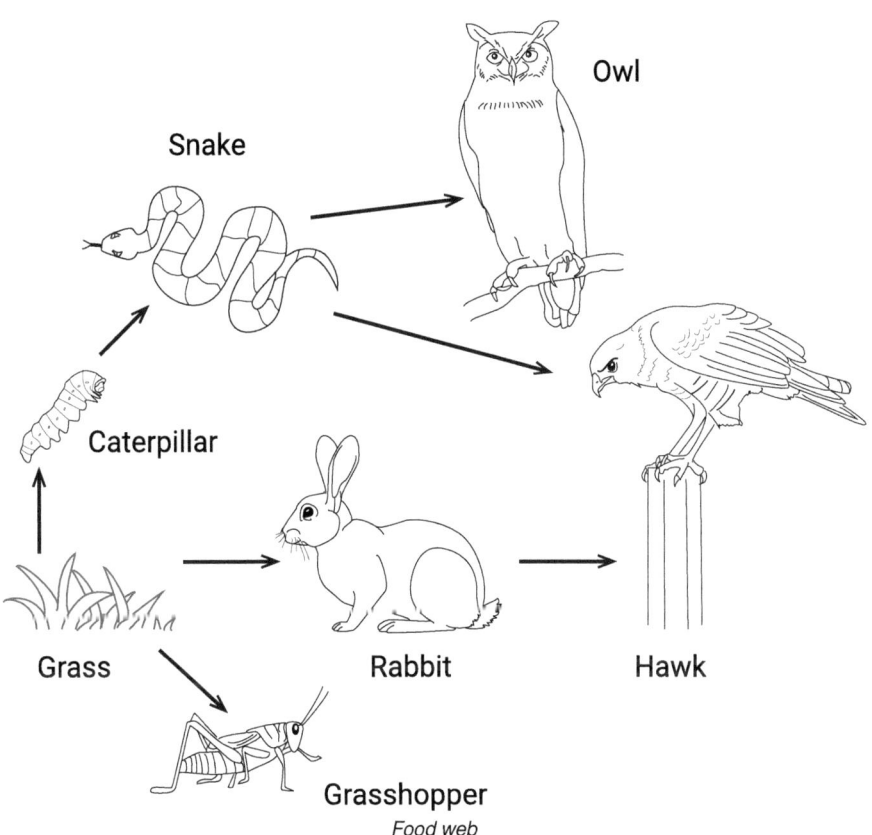
Food web

Summary of key ecological terms:

Term	Definition
Species	A group of organisms which can interbreed and produce fertile offspring.
Biodiversity	The range of species within an ecosystem.
Population	A group of organisms of the same species.
Producer	An organism which produces its own food.
Consumer	An organism which consumes other organisms for food.
Herbivore	An organism which only consumes plant matter.
Carnivore	An organism which only consumes animal matter.
Omnivore	An organism which consumes both plant and animal matter.
Predator	An animal which hunts other animals (prey) for food.
Prey	An organism which is hunted and killed by a predator.
Food chain	A diagram which shows a simple feeding relationship between organisms.
Food web	A diagram which shows the interconnection of food chains.
Ecosystem	Consists of all the organisms (the community) living in a particular habitat and the non-living components with which the organisms interact.

1.2 Interactions of organisms in food webs

A food web is a diagram which shows the detailed feeding relationships between organisms in an ecosystem. The arrows in a food web point from the food to the feeder and represent the flow of energy from one organism to another. Food webs always start with a producer, this is often a plant for terrestrial food webs and plant plankton for marine food webs. The diagram below shows one example of a marine food web.

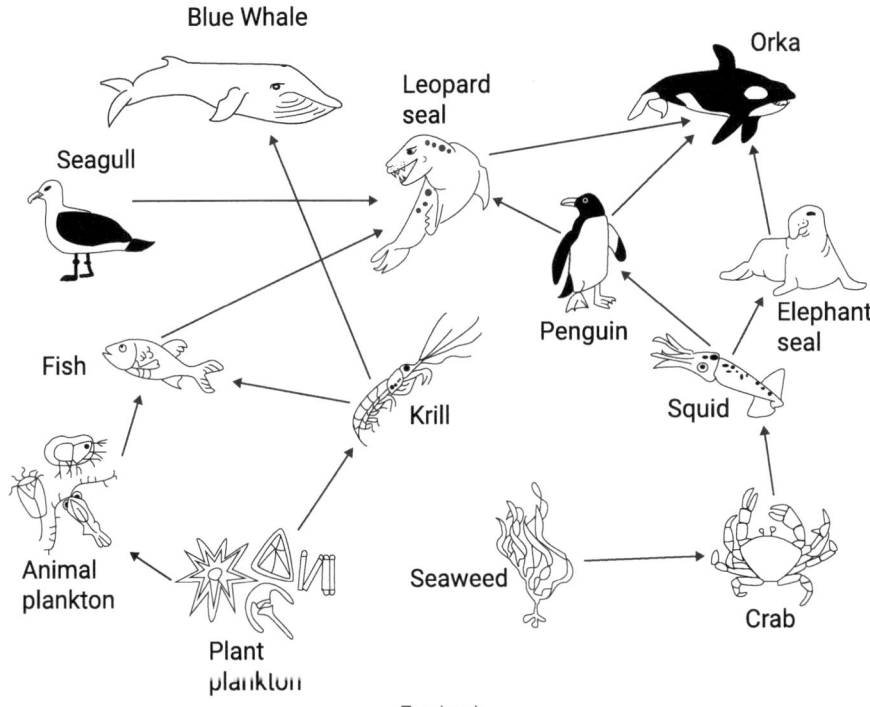
Food web

Food webs are delicately balanced and the removal of one organism can affect all other organisms in the food web. For example krill are a **crustacean** which is at risk of overfishing. If krill numbers drop below a critical level the numbers of blue whales may decrease due to a lack of food. A decrease in krill numbers can also affect other organisms in the food web, for example with less krill available fish may be forced to eat more animal plankton; therefore the loss of krill would be likely to reduce the fish population due to loss of a food source and reduce the animal plankton population due to increased predation. It is difficult to predict the effect of removing a species from a food web for example the loss of krill may cause a reduction in the animal plankton population as described above, however, it may also cause an increase in the animal plankton population as a result of increased plant plankton population and therefore food availability.

1.3 Niche

A niche is the role that an organism plays within a community. An organism's niche describes all of its interactions with the abiotic and biotic factors within its environment. Abiotic factors may relate to the resources an organism requires in terms of light intensity or the conditions an organism can tolerate such as temperature. Biotic factors may relate to an organism's interactions with other organisms in the community such as competition and predation.

The niche of a barn owl
Barn owls are found in many **temperate** climates but cannot live in polar regions due to the low winter temperatures. Barn owls are active at night when they hunt for food such as voles, mice and rats. They are capable of hunting in complete darkness and locate prey by hearing rather than sight. Barn owls nest in hollows found naturally in trees, cave or cliffs; they also make use of man-made structures such as nest boxes and barns. Barn owls are preyed upon by larger birds of prey such as buzzards. Barn owls compete with other owl species such as tawny owls for food.

Barn owl

Video: What is a niche? Go online

Watch this video which talks about an organism's niche.

https://www.youtube.com/watch?v=xlVixvcR4Jc

1.4 Competition

Competition occurs when the resources organisms require are in short supply. Animals compete for resources such as food, territory and mates. Plants compete for resources such as light, space and soil nutrients.

TOPIC 1. ECOSYSTEMS

Sea anemones compete for territory

Trees compete for light

Competition can be classed as interspecific or intraspecific. Interspecific competition occurs amongst individuals of different species for one or a few of the resources they require. Intraspecific competition occurs amongst individuals of the same species and is for all resources required. Intraspecific competition is therefore more intense than interspecific competition as members of the same species have the exact same requirements as each other.

Interspecific competition

Intraspecific competition

1.5 Learning points

Summary

- A species is a group of organisms which can interbreed and produce fertile offspring.
- Biodiversity refers to the range of species within an ecosystem.
- A population is a group of organisms of the same species.
- A producer is an organism which produces its own food.
- A consumer is an organism which consumes other organisms for food.
- A herbivore is an organism which only consumes plant matter.
- A carnivore is an organism which only consumes animal matter.
- An omnivore is an organism which consumes both plant and animal matter.
- A predator is an animal which hunts other animals (prey) for food.
- Prey is an organism which is hunted and killed by a predator.
- A food chain is a diagram which shows a simple feeding relationship between organisms.
- An ecosystem consists of all the organisms (the community) living in a particular habitat and the non-living components with which the organisms interact.
- A niche is the role that an organism plays within a community.
- A niche relates to the resources an organism requires in its ecosystem, such as light and nutrient availability and its interactions with other organisms in the community. A niche also involves competition and predation and the conditions an organism can tolerate such as temperature.
- Competition occurs when the resources organisms require are in short supply.
- Interspecific competition occurs amongst individuals of different species for one or a few of the resources they require.
- Intraspecific competition occurs amongst individuals of the same species and is for all resources required.
- Intraspecific competition is more intense than interspecific competition.

1.6 End of topic test

End of topic test: Ecosystems Go online

The diagram below shows a woodland food web.

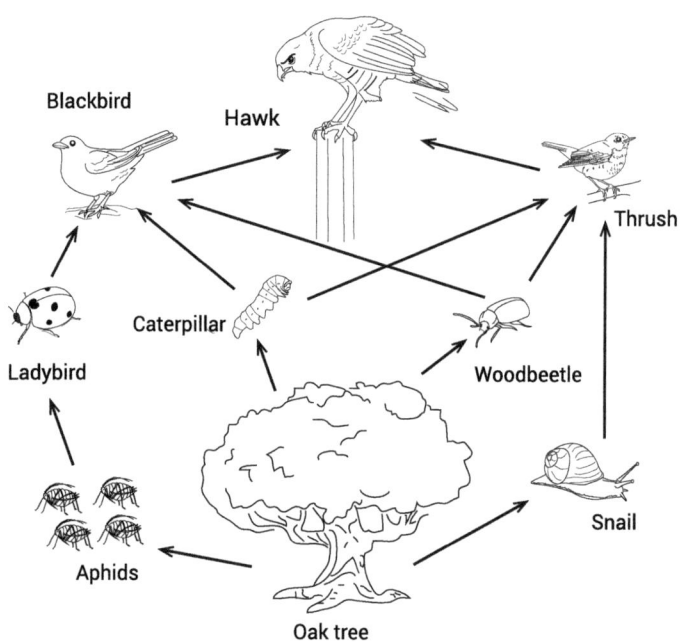

Q1: Using information from the food web give three ecological terms which could be used to describe the hawk.

....

Q2: What term describes organisms such as oak trees which are capable of making their own food?

....

Q3: Select organisms from the food web to complete the food chain below.

oak tree → _____ → _____ → _____

....

Q4: Name two organisms which are in competition with each other:

....

TOPIC 1. ECOSYSTEMS

Q5: Complete the following sentence:

If the aphids were killed by a pesticide the number of ladybirds would be likely to (increase/decrease/stay the same) due to [open text box].

Rabbits and hares are organisms which look very similar but they are in fact different species. Rabbits and hares have a similar role within the ecosystem they inhabit; both eat grasses, occupy similar habitats and are eaten by foxes.

Rabbit   Hare

Q6: Give a definition of the term species.

....

Q7: What term describes the role an organism plays within its ecosystem?

....

Q8: What type of competition occurs between rabbits and hares?

....

Q9: What term describes organisms such as rabbits and hares which are hunted, killed and eaten by predators such as foxes?

The picture below shows a pond ecosystem. Ponds contain delicately balanced food webs. In the pond below the pond weed is eaten by ducks and snails. Algae is eaten by snails, fish and tadpoles. Snails and fish in turn are eaten by ducks and frogs.

Q10: Complete the sentence below by selecting the correct option from each box to define the term ecosystem.

An ecosystem consists of all the organisms (the [population / community / habitat]) living in a particular [population / community / habitat] and the non-living components with which the organisms interact.

....

Q11:
Use the information in the passage to complete the following food web.

TOPIC 1. ECOSYSTEMS

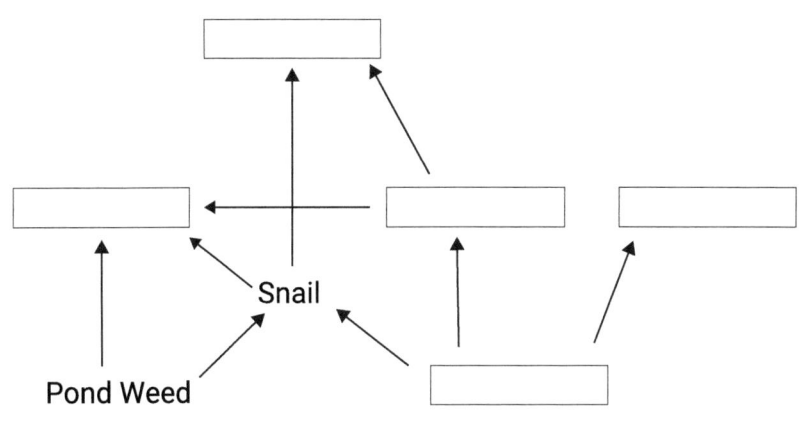

Q12: Complete the following sentence by selecting the correct option from the drop down box to describe the eating habits of ducks.

Ducks eat both plant and animal matter and are therefore described as [carnivores / herbivores / omnivores].

....

Q13: What would happen to the snail population if the ducks were killed due to overhunting? Give **two** reasons for your answer.

- increase
- decrease
- stay the same

Unit 3 Topic 2

Distribution of organisms

Contents
- 2.1 Abiotic and biotic factors 212
- 2.2 Measuring abiotic factors 213
- 2.3 Sampling of plants and animals 215
- 2.4 Paired-statement keys 217
- 2.5 Indicator species 219
- 2.6 Learning points 221
- 2.7 Extended response 222
- 2.8 End of topic test 223

Learning objective

By the end of this topic you should be able to:

- define the terms abiotic factors and biotic factors;
- give examples of abiotic and biotic factors;
- describe the effect of biotic and abiotic factors on biodiversity and the distribution of organisms;
- identify factors which can cause an increase or a decrease in biodiversity;
- describe how abiotic factors such as light intensity, soil moisture, pH and temperature are measured;
- identify possible sources of error when measuring abiotic factors and describe how to minimise them;
- describe how plants and animals can be sampled using quadrats and pitfall traps;
- evaluate the limitations of quadrats and pitfall traps and identify sources of error in their use;
- describe the need for representative sampling and adequate replication when sampling plants and animals;
- use and construct paired-statement keys to identify organisms;
- describe the role of indicator species.

2.1 Abiotic and biotic factors

Abiotic factors are non-living factors such as light intensity, moisture, pH and temperature. Biotic factors are living factors such as competition for resources, disease, food availability, grazing and predation.

Both abiotic and biotic factors can affect the **distribution** of organisms and biodiversity of an ecosystem. Abiotic factors such as moisture have a large impact on the distribution of animals. For example barnacles are animals which require regular submersion in water. The picture below shows the distribution of barnacles on a rocky shore. The barnacles are incapable of living further up the shore due to a lack of water which would cause them to dry out.

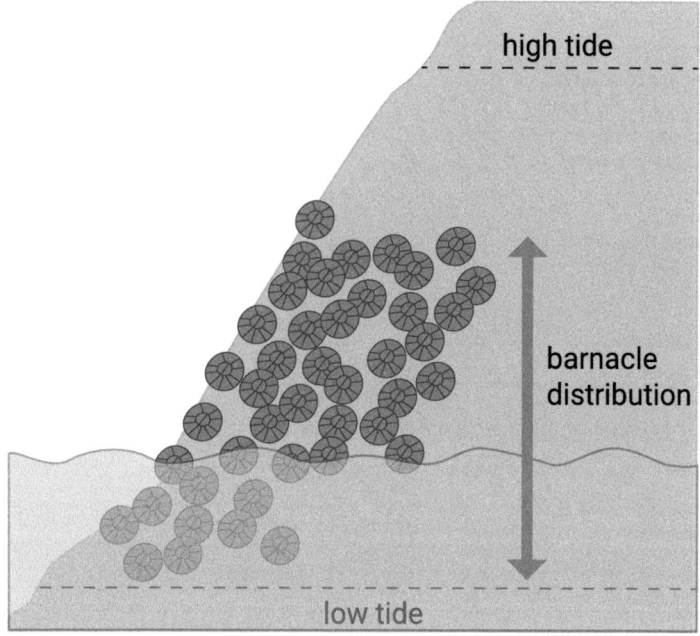

Barnacle distribution

Biotic factors also affect the distribution of organisms. Food availability is a key biotic factor for all consumer organisms. Any species of animal will only be found in areas of their ecosystem which present an adequate food supply.

Abiotic factors can affect biodiversity. In general warm temperatures increase biodiversity and extremely low or high temperatures decrease biodiversity. This can be demonstrated when comparing a warm tropical ecosystem such as the rainforest which has high biodiversity to an arctic or desert environment, both of which have comparatively lower levels of biodiversity.

Biotic factors such as grazing can also have a large impact on the biodiversity of an ecosystem. Low levels of grazing allow dominant grass species to grow uncontrollably, preventing more delicate species from growing thus decreasing biodiversity. High levels of grazing control the growth of dominant grass species allowing more delicate species to grow thus increasing biodiversity. At very high levels of grazing all species are eaten and biodiversity decreases.

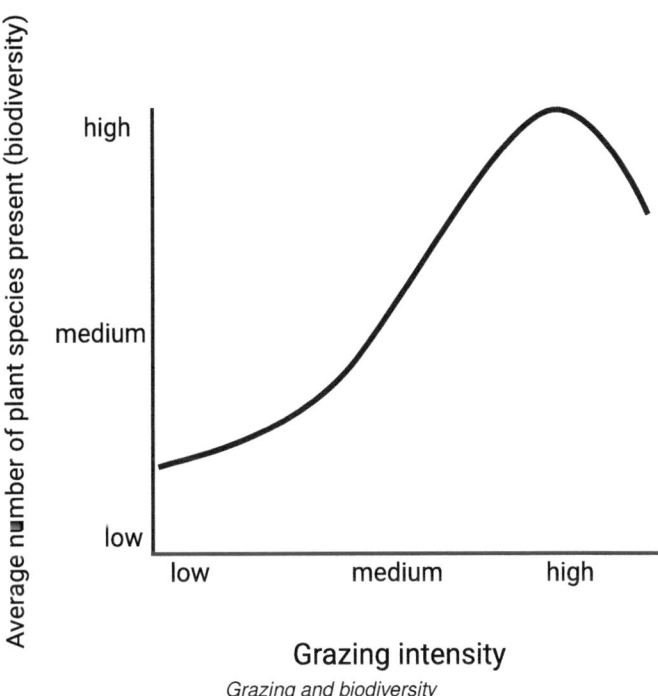

Grazing and biodiversity

2.2 Measuring abiotic factors

Light intensity is measured using a light meter. The sensor is held upward and a reading is taken from the scale on the meter. When using a light meter the user and other bystanders may inadvertently shade the sensor therefore it is important to ensue no shadows are cast over the light sensor.

Soil moisture is measured using a moisture meter. The probe is pushed into the ground and a reading is taken from the scale on the meter. When using a soil moisture meter it is important to ensure there is no moisture left on the probe from a previous reading therefore the probe should be wiped in between measurements. A pH meter is used to measure pH. This meter is used in the same way as a moisture meter. Similarly users should ensure that soil is not left on the probe in between readings to ensure an accurate pH value is obtained.

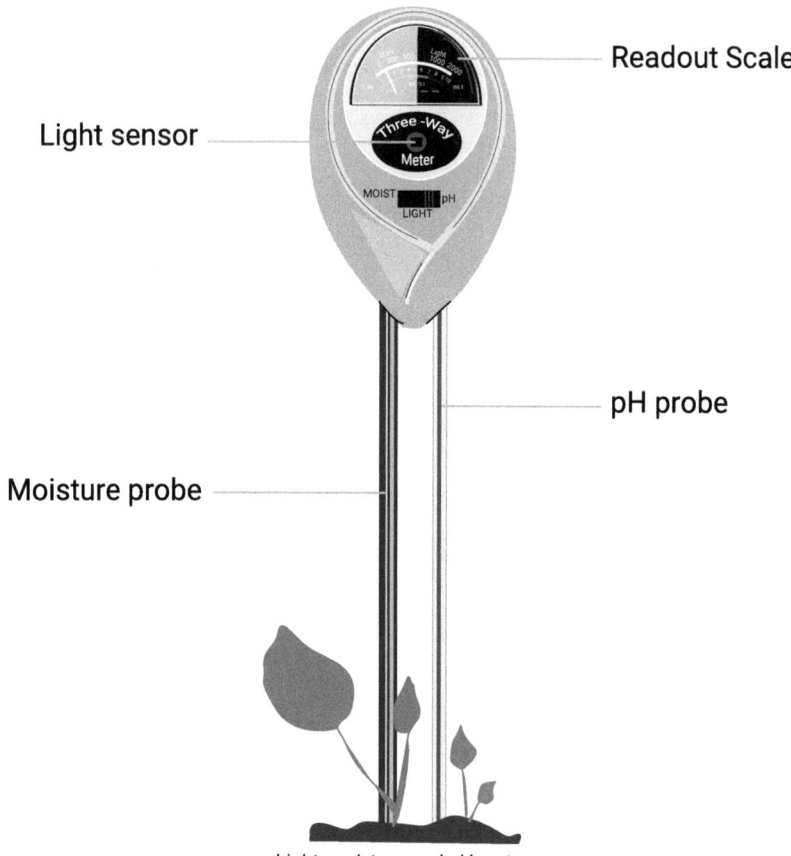
Light, moisture and pH meter

Temperature is measured using a thermometer. The thermometer is placed in the experimental area, allowed to stabilise and a reading is taken from the scale. When using a thermometer direct sunlight or heat from the user's hand can result in inaccurate readings, therefore it is important to ensure the thermometer is set up in the shade and not held by the user.

TOPIC 2. DISTRIBUTION OF ORGANISMS

Thermometer

2.3 Sampling of plants and animals

Quadrats are used for estimating the abundance of plants or slow moving animals (such as limpets) in an ecosystem. The quadrat is thrown at random and the number of squares that contain the investigated organism are counted. To be representative of the total population an adequate number of quadrats are randomly thrown. An average number of each organism per quadrat is calculated then scaled up represent the area.

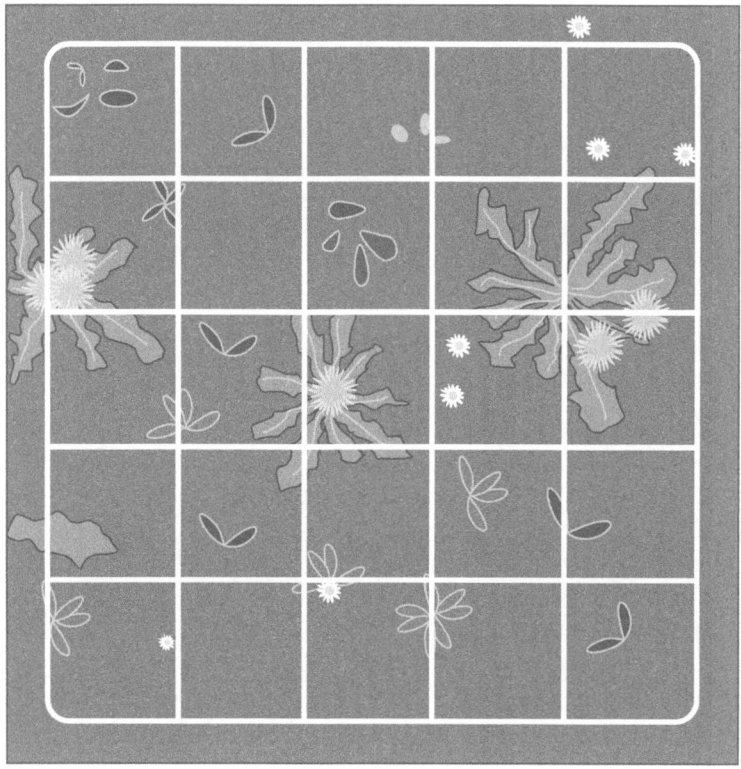

Quadrat

One source of error which can arise when using quadrats is determining whether an individual organism is in a certain square or not, for example an organism on the outer edge of the quadrat. To overcome this issue scientists set up and follow certain rules when using quadrats, for example borderline organisms on two specified sides being counted as 'in' and on the other two sides 'out'. Quadrats only sample a small percentage of the ecosystem as a whole therefore to ensure the results are representative the quadrat should be thrown randomly many times and an average calculated.

Pitfall traps are used to sample invertebrates which live on the ground. A hole is dug into the ground and a plastic pot is placed inside. It is important that the trap is level with the ground to ensure the invertebrates can fall in. The trap is camouflaged and left for a set period of time after which the number of each invertebrate found within the trap is counted.

TOPIC 2. DISTRIBUTION OF ORGANISMS

Pitfall trap

When setting up pitfall traps there are several potential issues:

- water getting in when it rains;
- birds taking trapped animals;
- predatory animals like spiders, ground beetles or centipedes eating their fellow captives.

These issues can be overcome by:

- putting a few pin pricks in the bottom to allow drainage;
- covering the trap with leaves to camouflage it;
- checking the trap often.

When sampling invertebrates using pitfall traps it is important to set several traps across the investigated area and calculate averages to ensure the results are representative of the area being investigated.

2.4 Paired-statement keys

Biological keys are used to identify organisms. A paired statement key is a list of statements which describe certain observable features of organisms. To use a paired statement key the user simply reads the descriptions in each pair and determines which applies to the organism they are trying to identify. This process is repeated down the list of statements until the name of the organism is determined. The paired statement key below can be used to identify leaf types.

Leaf A

Leaf B

1. Leaf made up of one part: go to 2.
   Leaf made up of several smaller parts: go to 3.
2. Outer leaf edge is wavy: **Oak**.
   Outer leaf edge is smooth: go to 4.
3. Leaflets attached at a single point on the leaf stalk: **Horse Chestnut**.
   Leaflets attached in pairs on the leaf stalk: **Rowan**.
4. Leaf is as wide as it is long: **Beech**.
   Leaf is longer than it is wide: **Laurel**.

To identify leaf A from pair number 1 the statement which applies is "Leaf made up of several smaller leaflets" therefore pair number 3 is considered next. From pair number 3 the statement which applies is "Leaflets attached at a single point on the leaf stalk" therefore leaf A is a horse chestnut leaf.

To identify leaf B from pair number 1 the statement which applies is "Leaf made up of one part" therefore pair number 2 is considered next. From pair number 2 the statement which applies is "Outer leaf edge is smooth" therefore pair number 4 is considered next. From pair number 4 the statement which applies is "Leaf is longer than it is wide" therefore leaf B is a laurel leaf.

TOPIC 2. DISTRIBUTION OF ORGANISMS

2.5 Indicator species

Indicator species are species that by their presence or absence indicate environmental quality or levels of pollution. Invertebrates are often used as indicator species to gauge levels of pollution in freshwater ecosystems. Organisms such as stonefly nymphs are very sensitive to pollution and therefore are abundant in water with low levels of pollution but absent from water with high levels of pollution. Organisms such as rat-tailed maggots are tolerant of high levels of pollution and therefore are abundant in water with high levels of pollution.

Stonefly nymph

Rat-tailed maggot

Lichen are often used as indicators of air pollution. Some types of lichen such as shrubby lichens are particularly sensitive to sulfur dioxide and are not capable of growing in areas with high levels of air pollution. Other species such as crusty lichens are more tolerant of moderate levels of sulfur dioxide and are therefore capable of growing in areas with moderate levels of air pollution.

Shrubby lichen

Crusty lichen

2.6 Learning points

> **Summary**
>
> - Biotic factors are living factors and abiotic factors are non-living factors.
> - Competition for resources, disease, food availability, grazing and predation are biotic factors.
> - Light intensity, moisture, pH and temperature are abiotic factors.
> - Both biotic and abiotic factors can affect the biodiversity and the distribution of organisms.
> - Some factors such as warm temperatures increase the biodiversity of an ecosystem while others such a low levels of grazing decrease the biodiversity of an ecosystem.
> - Light intensity is measured using a light meter, the sensor is held upwards and a reading is taken from the scale. The light sensor may be inadvertently shaded so it is important to ensure all users stand clear.
> - Soil moisture and pH are measured using a moisture meter or pH meter, the probe is inserted into the ground and a reading is taken from the scale. The probe on the meter may retain soil from a previous sample so it is important to wipe the probe in between readings.
> - Temperature is measured using a thermometer, the thermometer is placed in an open area and a reading is taken from the scale. Placing the thermometer in direct sunlight or holding it in a warm hand may provide an inaccurate reading so it is important to place the thermometer in a shaded area.
> - Quadrats are thrown randomly and the number of squares containing the organism of interest are counted. Quadrats only sample a small proportion of the investigated area therefore it is important to throw the quadrat randomly, several times and calculate an average.
> - Pitfall traps are sunk into the ground and camouflaged to allow invertebrates to fall in. It is important to ensure the trap is level with the soil and camouflaged so insects will fall in.
> - Paired statement keys use a series of statements to identify organisms.
> - Indicator species are species that by their presence or absence indicate environmental quality/levels of pollution.

2.7 Extended response

Extended response

Q1:

Describe how **ONE** of the following abiotic factors are measured:

- light intensity
- soil moisture
- soil pH
- temperature

Identify one possible source of error when measuring this factor and describe how to minimise it.

4 marks

2.8 End of topic test

End of topic test: Distribution of organisms   Go online

The following list identifies factors which can affect the distribution of organisms in an ecosystem.

- Light intensity
- Temperature
- Predation
- Competition
- Soil pH
- Disease

Q2: Complete the following table to identify whether each factor is abiotic or biotic.

Abiotic factors	Biotic factors

....

For each of the following factors, name the instrument used to measure it and describe how the instrument is used.

Q3:
Light intensity
Instrument: _____
Description of use: _____

....

Q4:
Temperature
Instrument: _____
Description of use: _____

....

Q5:
Soil pH
Instrument: _____
Description of use: _____

....

The following table shows information about four different types of British butterfly.

Species	Main wing colour	Wingspan (mm)	Eyespots
Small white	white	35-45	present
Gatekeeper	orange	35-45	present
Common blue	blue	30-34	absent
Brimstone	yellow	65-75	absent
Small heath	orange	29-33	present

Use the information in the table above to complete the following paired statement key.

Q6:

1. Wingspan greater than 35mm: go to ___.
   Wingspan less than 35 mm: go to___.
2. _____: go to 4
   _____: go to Brimstone
3. Main wing colour blue: Common blue
   Main wing colour orange: Small heath
4. Main wing colour white: _____
   Main wing colour orange: _____

Two groups of students used quadrats to determine the biodiversity of an area of grassland.

Group 1 results

Plant species	Abundance
Daisy	6
Buttercup	8
Dandelion	11
Plantain	7

Group 2 results

Plant species	Abundance Quadrat 1	Abundance Quadrat 2	Abundance Quadrat 3	Abundance Average
Daisy	7	5	9	7
Buttercup	4	5	3	4
Dandelion	12	8	16	12
Plantain	4	6	2	4

Q7: Describe how to use a quadrat.

....

Q8: Explain why the results from group 2 are more reliable than the results from group 1.

....

Q9: The students also investigated the biodiversity of invertebrates living in the grassland. Name a sampling technique they could have used.

....

Q10: The biodiversity of key insects can reveal information about the levels of pollution in an ecosystem. What name is given to organisms which by their presence or absence indicate environmental quality?

The following key can be used to identify invertebrates found in freshwater.

1. Legs present: go to 2
   Legs not present: go to 3

2. Eight legs: **water mite**
   More than eight legs: **sowbug**

3. Shell present: **water snail**
   No shell present: go to 4

4. Segmented body: **leech**
   Non-segmented body: **flatworm**

Q11: Use the key to identify the following invertebrate.

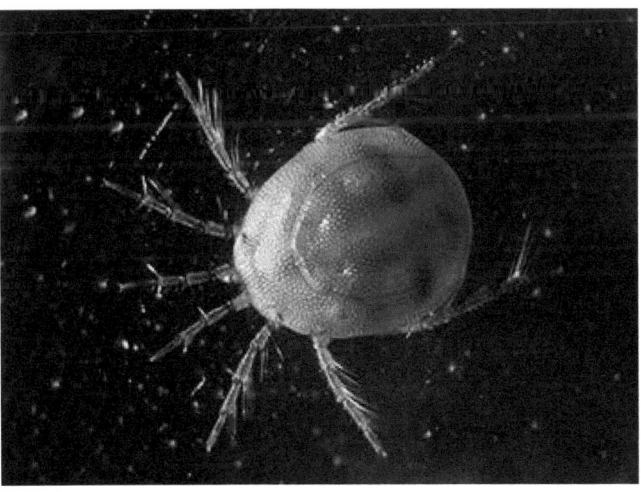

....

Q12: Use the key to identify the following invertebrate.

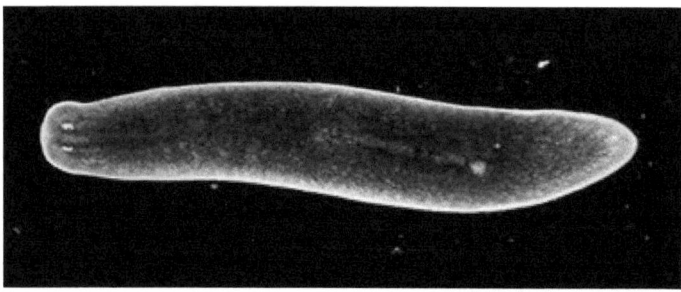

Unit 3 Topic 3

Photosynthesis

Contents

3.1 Reactions of photosynthesis 228
3.2 Limiting factors 229
3.3 Learning points 232
3.4 Extended response 232
3.5 Extension materials 233
3.6 End of topic test 235

Learning objective

By the end of this topic you should be able to:

- state that photosynthesis is a two-stage process;
- describe the first stage of photosynthesis (light reactions);
- describe the second stage of photosynthesis (carbon fixation);
- give the word summary of the process of photosynthesis;
- state that the chemical energy in sugar is available for respiration or the sugar can be converted into other substances, such as starch (storage) and cellulose (structural);
- name three limiting factors of photosynthesis;
- describe the impact of limiting factors on photosynthesis and plant growth;
- analyse limiting factors graphs.

3.1 Reactions of photosynthesis

Photosynthesis is a reaction which converts light energy into chemical energy. During photosynthesis light energy along with carbon dioxide and water are used to produce sugar and oxygen.

Word summary of the process of photosynthesis:

$$\text{carbon dioxide + water} \xrightarrow{\text{light energy}} \text{sugar + oxygen}$$

Photosynthesis is a two-stage process, the first stage is called the light reactions and the second stage is called carbon fixation.

During the light reactions light energy from the sun is trapped by a green pigment called chlorophyll in the chloroplast. The light energy is converted into chemical energy which is used to generate ATP. Water is split to produce hydrogen and oxygen during the light reactions. The hydrogen is used in the second stage and the oxygen diffuses from the cell.

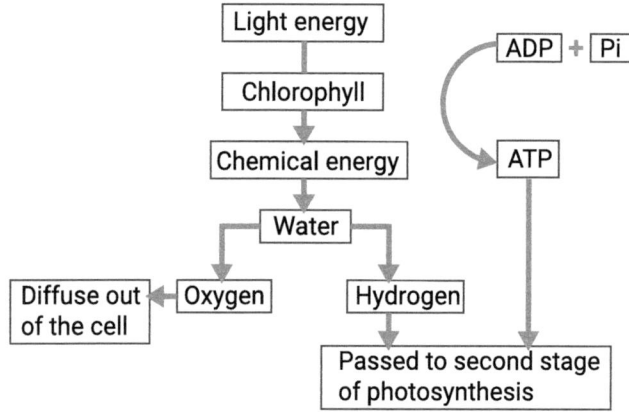

The light reactions

The second stage of photosynthesis, carbon fixation, uses hydrogen and ATP from the light reactions to convert carbon dioxide from the air into sugar. The carbon fixation reactions are catalysed by enzymes; this means that the rate of the carbon fixation reactions are affected by temperature.

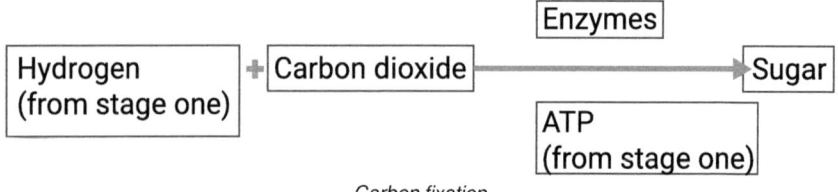

Carbon fixation

The sugar produced during photosynthesis can be used in respiration to provide energy for the cell.

The sugar produced by photosynthesis can also be converted into glucose molecules which can then be linked together to form the carbohydrate starch. Starch is a storage carbohydrate. Glucose molecules can also be linked together in a different way to produce the structural carbohydrate cellulose which forms part of the plant cell wall.

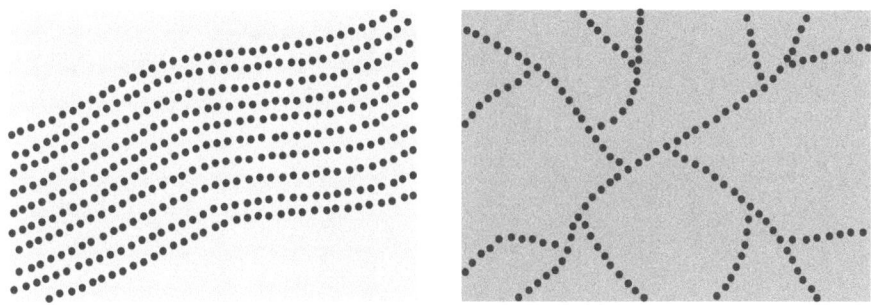

Cellulose (left) and starch (right)

3.2 Limiting factors

A limiting factor is any factor which limits the rate of photosynthesis when it is in short supply. The three limiting factors of photosynthesis are carbon dioxide concentration, light intensity and temperature. When the rate of photosynthesis is reduced, less sugar is produced, therefore plant growth is reduced also.

The rate of photosynthesis can be monitored by measuring the volume of oxygen released or the mass of the plant. Experiments can be conducted to investigate the effect of limiting factors on the rate of photosynthesis. The results can be presented in a graph and analysed.

Both light intensity and carbon dioxide concentration affect the rate of photosynthesis in a similar manner. As either factor is increased the rate of photosynthesis will also increase until a certain point, after which the rate of photosynthesis will remain constant.

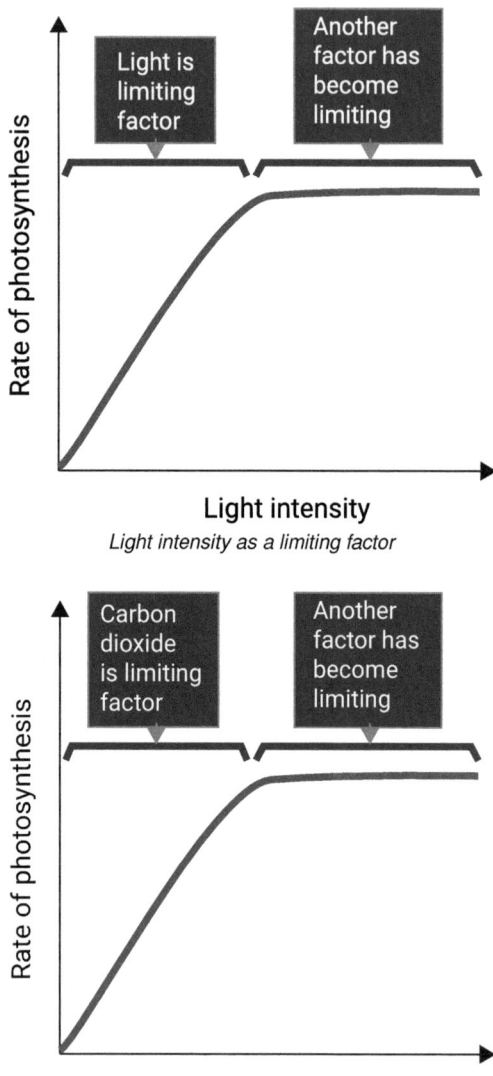

Light intensity as a limiting factor

Carbon dioxide concentration as a limiting factor

Temperature affects the rate of photosynthesis in a different manner compared to light intensity and carbon dioxide concentration. Carbon fixation is a series of enzyme controlled reactions therefore photosynthesis is heavily temperature dependant. As the temperature increases so does the rate of photosynthesis until an optimum, after which the rate of photosynthesis decreases as the high temperature begins to denature the enzymes which control the carbon fixation stage of photosynthesis.

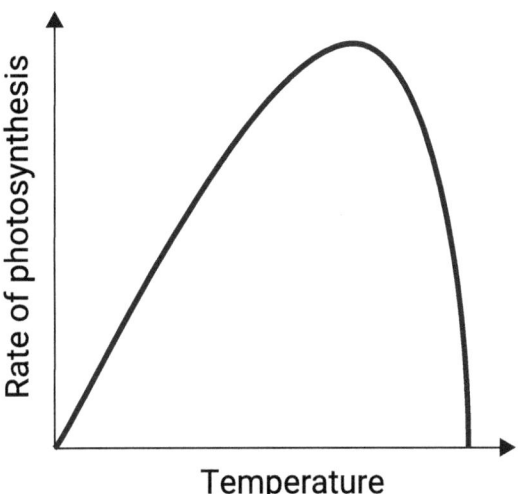

Temperature as a limiting factor

Experiments can be conducted to investigate the effect of multiple limiting factors simultaneously. The following graph shows the results of an experiment which investigated the effect of both light intensity and carbon dioxide concentration on the rate of photosynthesis.

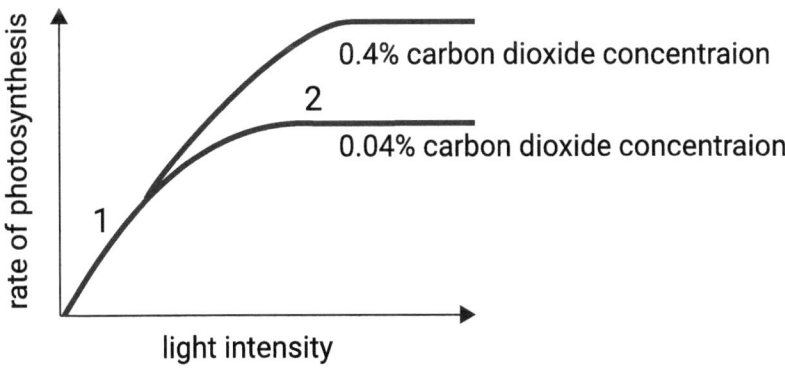

Limiting factors experiment

At point 1 on the graph light intensity is the factor which is limiting the rate of photosynthesis. This can be determined as when the light intensity increases further the rate of photosynthesis also increases. At point 2 on the graph light is no longer the limiting factor. This can be determined as when the light intensity increases further the rate of photosynthesis remains constant (it does not increase further). This means another factor is limiting the rate of photosynthesis, in this case carbon dioxide concentration. This can be determined as when the carbon dioxide concentration is increased (from 0.04% to 0.4%) the rate of photosynthesis increases.

3.3 Learning points

Summary

- Photosynthesis is a two-stage process.
- The first stage of photosynthesis is called the light reactions.
- During the light reactions:
  - the light energy from the sun is trapped by chlorophyll in the chloroplasts and is converted into chemical energy which is used to generate ATP;
  - water is split to produce hydrogen and oxygen;
  - oxygen diffuses from the cell.
- The second stage of photosynthesis is called carbon fixation.
- Carbon fixation is a series of enzyme-controlled reactions.
- During carbon fixation hydrogen and ATP (produced by the light reactions) with carbon dioxide are used to produce sugar.
- Word summary of the process of photosynthesis:

$$\text{carbon dioxide + water} \xrightarrow{\text{light energy}} \text{sugar + oxygen}$$

- The chemical energy in sugar is available for respiration or the sugar can be converted into other substances, such as starch (storage) and cellulose (structural).
- The limiting factors of photosynthesis are carbon dioxide concentration, light intensity and temperature.
- If any of the limiting factors of photosynthesis are in low supply or absent the rate of photosynthesis will slow down and so too will plant growth.
- The effects of the limiting factors on the rate of photosynthesis can be investigated and the results plotted onto a graph to allow analysis.

3.4 Extended response

Extended response

Q1: Describe the light reactions stage of photosynthesis.

4 marks

....

Q2: Describe the carbon fixation stage of photosynthesis.

4 marks

3.5 Extension materials

The importance of photosynthesis

The process of photosynthesis shaped the evolution of life on Earth billions of years ago and its importance continues to this day.

Although photosynthesis is most commonly linked with green plants, the first photosynthetic cells were probably a type of bacteria called cyanobacteria. Billions of years ago Earth's atmosphere held a very different balance of gases compared to today; it had much higher levels of carbon dioxide and very little oxygen. With the evolution of photosynthesising cyanobacteria the oxygen levels began to increase and this paved the way for the evolution of the wide variety of aerobic (oxygen-using) organisms we see present on Earth today.

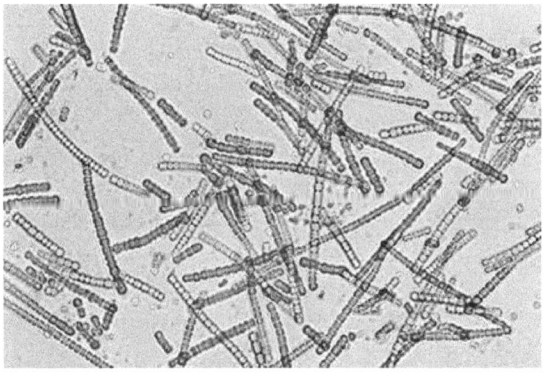

Cyanobacteria

Plants living millions of years ago carried out photosynthesis to produce food. The remains of these plants, along with other small animals which fed on them, were buried under layers of sand and silt. Over millions of years and as a result of exposure to extreme heat and pressure these remains were turned into fossil fuels (coal, oil and gas). The production of these vitally important raw materials would not have been possible without the process of photosynthesis.

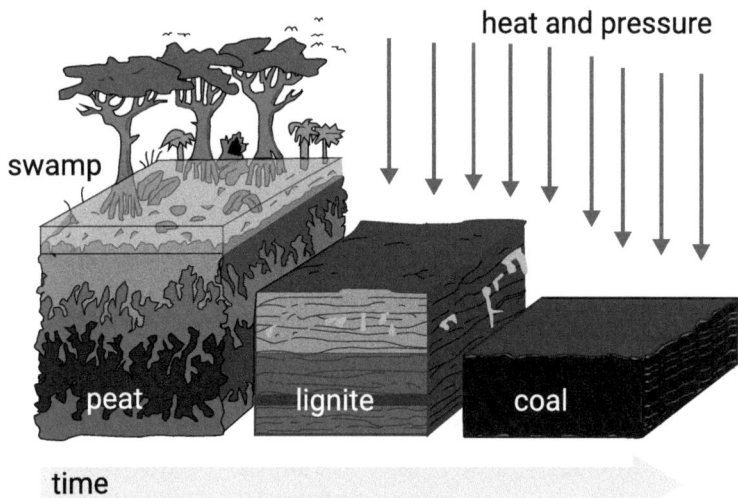

Formation of coal

Humans rely on photosynthesis for a wide range of resources. Every food web on Earth begins with a producer, an organism capable of making its own food. Most commonly the producer is a photosynthesising organism such as a green plant or algae. Plants (and therefore photosynthesis) also provide humans with a wide range of raw materials and medicines as well as maintaining the gas balance in the atmosphere.

3.6 End of topic test

End of topic test: Photosynthesis — Go online

Green plants produce food by a two stage process called photosynthesis. The diagram below shows a cell taken from a green plant.

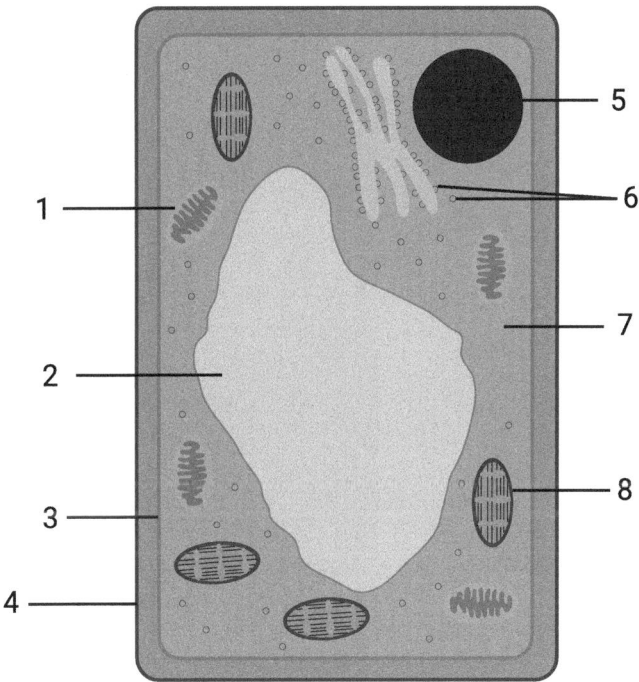

Q3: Name the first stage of photosynthesis.

....

Q4: Using a number from the diagram, identify the organelle where photosynthesis takes place.

....

Q5: Describe the role of chlorophyll in photosynthesis.

....

Q6: Name one product which is produced in the first stage and required for the second stage of photosynthesis.

....

Q7: Name the second stage of photosynthesis.

....

Q8:

The diagram below represents the second stage of photosynthesis.

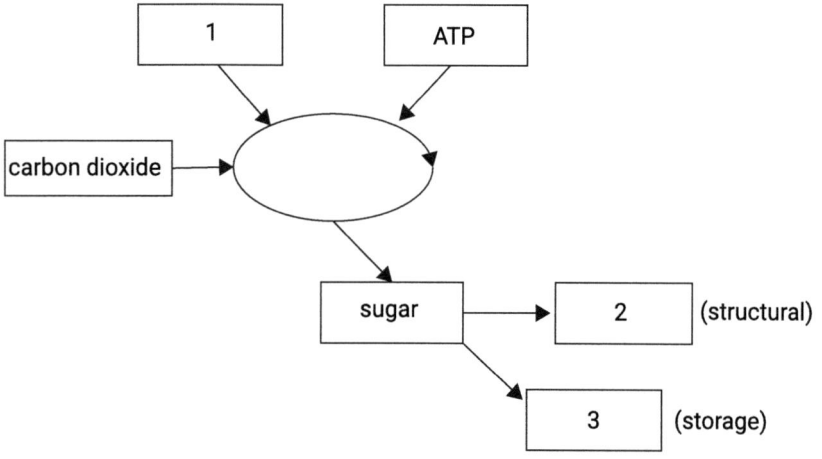

Name substances 1, 2 and 3.

....

Q9: Explain why photosynthesis cannot occur at extremely high temperatures.

An experiment was conducted to investigate the effect of carbon dioxide concentration on the rate of photosynthesis at different temperatures. The results are shown in the graph below.

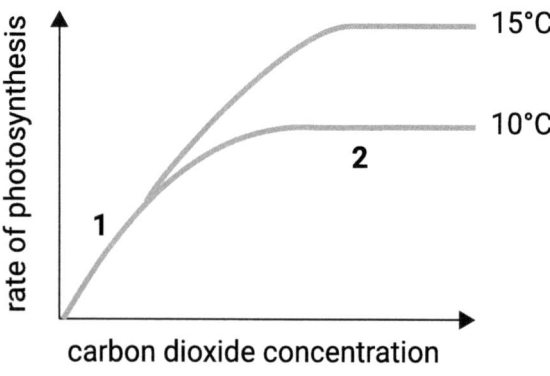

Q10: Other than carbon dioxide concentration and temperature name one limiting factor of photosynthesis.

....

Q11: State the factor which is limiting the rate of photosynthesis at point 1 on the graph.

....

Q12: State the factor which is limiting the rate of photosynthesis at point 2 on the graph.

Unit 3 Topic 4

Energy in ecosystems

Contents

4.1 Energy in ecosystems 240
4.2 Learning points 242
4.3 End of topic test 243

Learning objective

By the end of this topic you should be able to:

- state that the majority of the energy available at one level of a food chain is not passed on to the next level;
- name three ways energy is lost from a food chain;
- state that only a very small quantity of energy is used for growth and is therefore available at the next level in a food chain;
- define the terms pyramid of numbers and pyramid of energy;
- state that irregular shapes of pyramids of numbers based on different body sizes can be represented as true pyramids of energy.

4.1 Energy in ecosystems

Food chains show feeding relationships within an ecosystem. The arrows in the food chain represent the flow of energy from food to feeder, however, not all energy is passed from one level of the food chain to the next. Only a very small quantity of energy (around 10%) is used for growth and is therefore available at the next level in a food chain. The remaining energy (around 90%) is lost as heat, movement or undigested materials.

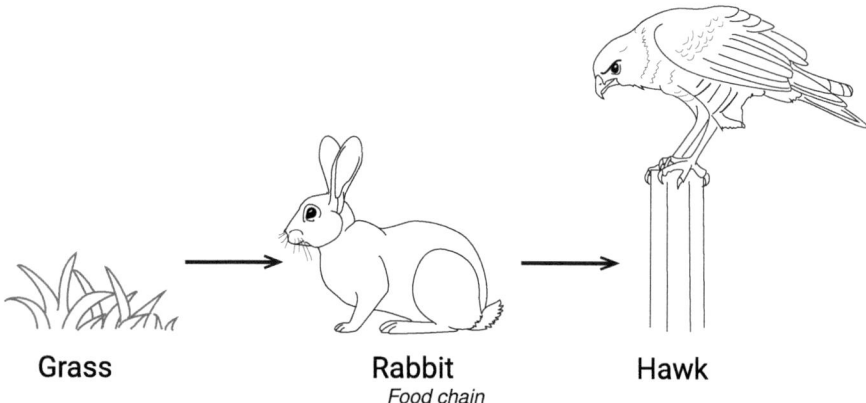

Food chain

A pyramid of numbers shows the number of organisms at each level of a food chain. Each level of the pyramid represents the number of organisms at that level of the food chain. The larger the bar the more organisms are present at that level of the food chain.

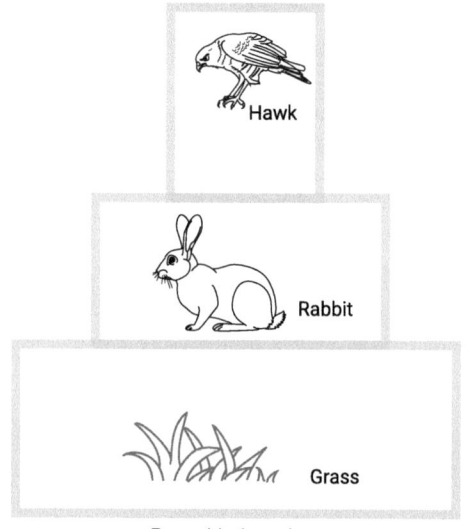

Pyramid of numbers

TOPIC 4. ENERGY IN ECOSYSTEMS

A pyramid of numbers does not always produce a regular pyramid shape especially when the body size of the next organism in the food chain is smaller than the previous organism. For example the food chain shown below results in an irregularly shaped pyramid of numbers because one rosebush can provide food for many hundreds of greenfly.

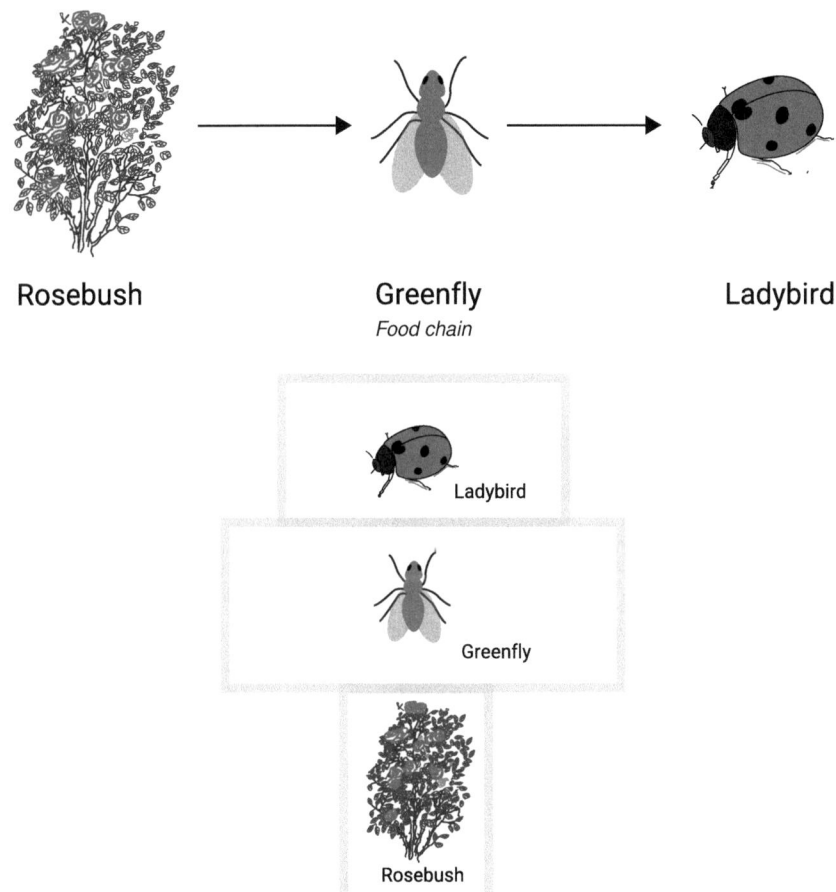

Irregular shapes of pyramids of numbers based on different body sizes can be represented as true pyramids of energy. A pyramid of energy shows the total quantity of energy stored in the **biomass** of organisms at each level of a food chain.

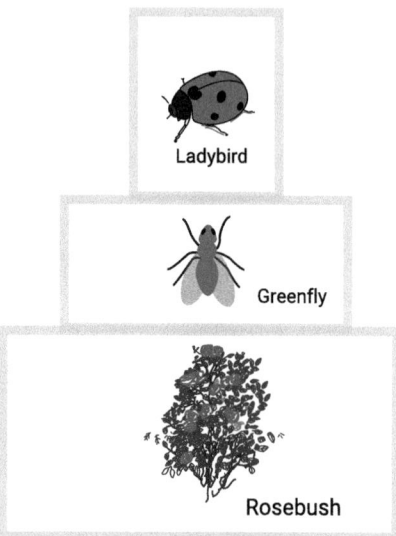

Pyramid of energy

4.2 Learning points

Summary

- In transfers from one level to the next in a food chain, the majority of the energy is lost as heat, movement or undigested materials.
- Only a very small quantity is used for growth and is therefore available at the next level in a food chain.
- A pyramid of numbers shows the number of organisms at each level of a food chain.
- A pyramid of energy shows the total quantity of energy stored in the biomass of organisms at each level of a food chain.
- Irregular shapes of pyramids of numbers based on different body sizes can be represented as true pyramids of energy.

4.3 End of topic test

End of topic test: Energy in ecosystems Go online

The diagram below shows a pyramid of numbers.

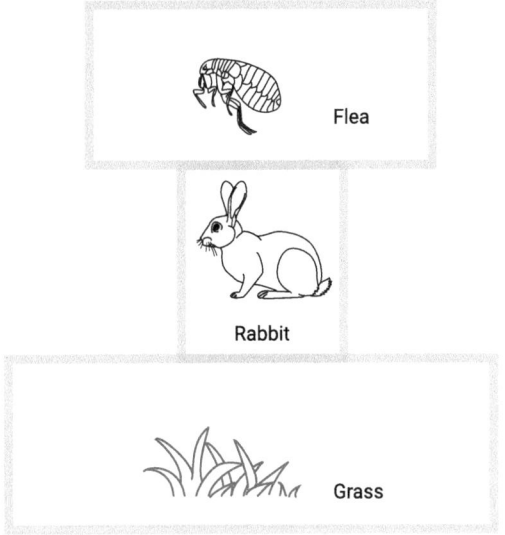

Q1: Give the meaning if the term pyramid of numbers.
....

Q2: The pyramid above shows an irregular shape. What type of diagram should be used to give a true pyramidal shape?

Only a very small quantity of energy is passed from one level of a food chain to the next

Q3: Name the process which allows energy to pass from one level of a food chain to the next.
....

Q4: Name one way energy is lost from a food chain.

Unit 3 Topic 5

Food production

Contents

5.1 The increasing human population 246
5.2 Fertilisers 246
5.3 Pesticides 248
5.4 Learning points 252
5.5 Extended response 253
5.6 Extension materials 253
5.7 End of topic test 255

Learning objective

By the end of this topic you should be able to:

- state that the increasing human population requires an increased food yield and this can involve the use of fertilisers and pesticides;

- describe the role of fertilisers;

- describe the role of pesticides;

- describe the role of nitrates in plants;

- describe the negative effects of fertilisers on the environment;

- state that genetically modified (GM) crops can be used to reduce the use of fertilisers;

- describe the negative effects of pesticides on non-target organisms;

- describe the use of biological control and genetically modified (GM) crops as alternatives to the use of pesticides.

5.1 The increasing human population

Over the last 200 years the human population has increased **exponentially** from 1 billion in 1800 to over 7 billion in 2018. The population is predicted to continue increasing, potentially reaching 11 billion by 2100. The increasing human population has put a lot of pressure on food production systems to meet the increased demand for food. This has resulted in farmers adopting intensive farming practises such as the use of fertilisers and pesticides to increase crop yields.

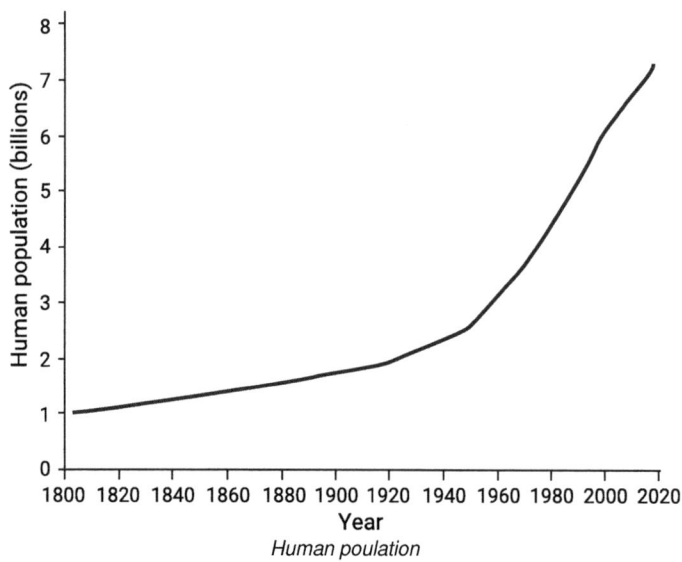

Human poulation

5.2 Fertilisers

Fertilisers provide chemicals such as nitrates which increase crop yield. Nitrates are a vitally important plant nutrient, they are used to produce amino acids which are synthesised into plant proteins. Nitrates are dissolved in soil water and absorbed into plants by their roots. Fertilisers can be added to soil to increase the nitrate content of the soil. Animals consume plants or other animals to obtain amino acids for protein synthesis.

Although fertilisers allow farmers to increase crop yield, they can have a negative impact on the environment. Fertilisers can leach into fresh water, adding extra, unwanted nitrates. This will increase algal populations which can cause algal blooms.

Algal bloom

Algal blooms reduce light levels, preventing aquatic plants from photosynthesising causing them to die. These dead plants, as well as dead algae, become food for bacteria which increase greatly in number. The bacteria use up large quantities of oxygen, reducing the oxygen availability for other organisms. When oxygen availability drops below a certain critical level the water can no longer support life meaning fish and other organisms die.

248 UNIT 3. LIFE ON EARTH

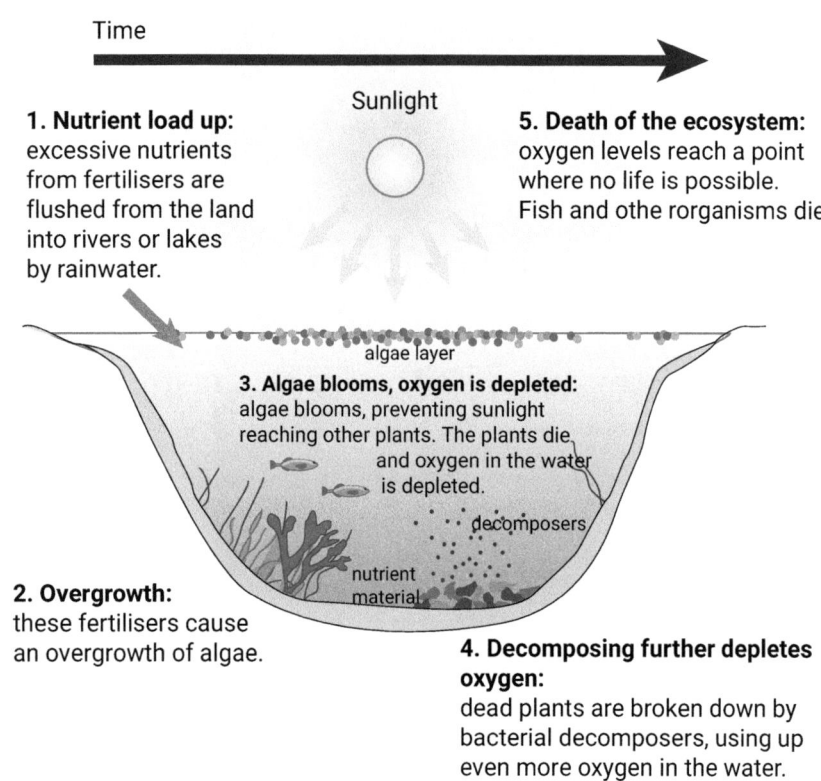

Negative effects of fertilisers

Scientists have been using genetic modification (GM) techniques to alter the DNA of crop plants to allow them to grow well with less fertilisers. For example a group of scientists have produced a genetically modified rice plant which can take up and use nitrogen from the soil more efficiently than varieties which are currently being grown. This scientific advancement allows farmers to reduce the quantities of fertiliser they apply while maintaining the crop yield.

5.3 Pesticides

Plants and animals which reduce crop yield are known as pests. Pests can be killed by pesticides for example herbicides are used to kill plant pests (e.g. weeds such as dandelion) and insecticides are used to kill insect pests (e.g. greenfly).

Crop pest

Pesticides are a useful tool for farmers as they allow the crop to grow to its full potential without damage (from insect pests), competition (from plant pests) or disease (from fungal pests). However, pesticides can have a negative impact on the environment. Pesticides sprayed onto crops can accumulate in the bodies of organisms over time; this means the concentration of the pesticide builds up in the tissues of the organism and can cause harm. This build-up of toxic substances in living organisms is known as bioaccumulation.

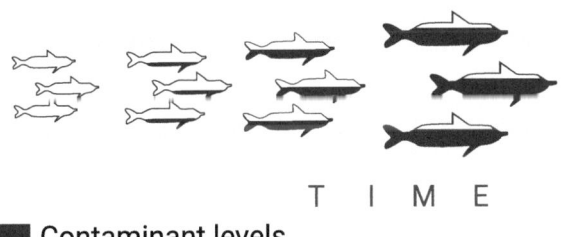

■ Contaminant levels

Bioaccumulation

As pesticides are passed along food chains, toxicity increases and can reach lethal levels. This build-up of toxic substances from one organism to another along a food chain is known as biomagnification.

■ Contaminant levels

Biomagnification

As a result of issues such as bioaccumulation, scientists are investigating alternative strategies to reduce the impact of pests without the use of pesticides.

Genetically modified (GM) crops offer one alternative to the use of pesticides. GM crops can be produced which do not require pesticide application, for example some varieties of GM cotton. Cotton plants are susceptible to attack from insect larvae pests such as cotton bollworms. Scientists have engineered GM cotton which produces Bt toxin in its tissues. Bt toxin kills pests such as the cotton bollworm and because the plant makes the toxin in its tissues there is no need to apply pesticides.

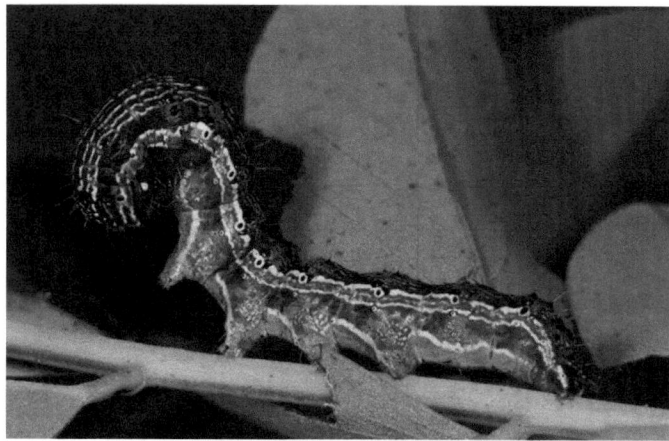

Cotton bollworm

Another alternative to the use of pesticides is biological control. Biological control involves controlling pest numbers by introducing a natural predator. For example the **parasitoid** wasp Aleiodes indiscretus can be used to control numbers of moth caterpillars. The wasp lays its eggs in the caterpillar, when the larvae hatch they use the caterpillar as a food source and in doing so kill it.

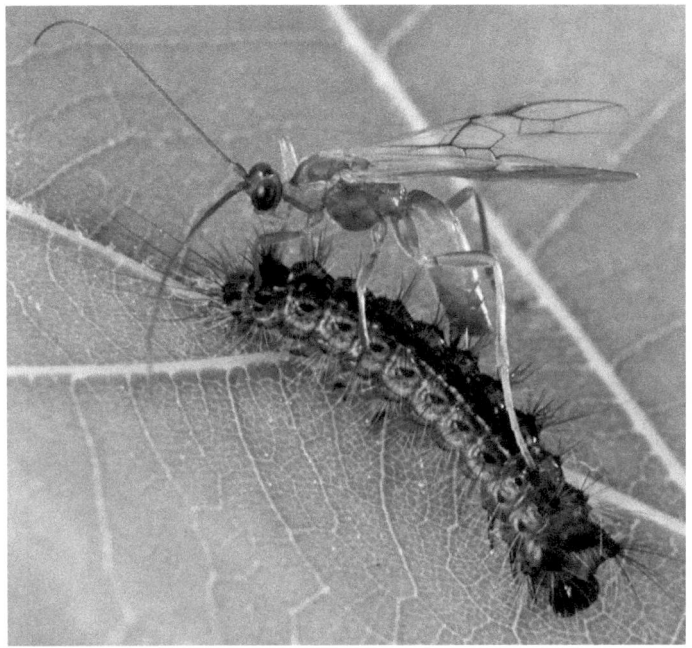
Biological control

5.4 Learning points

Summary

- An increasing human population requires an increased food yield.
- Increasing food production can involve the use of fertilisers and pesticides.
- Fertilisers provide chemicals such as nitrates which increase crop yield.
- Plants and animals which reduce crop yield can be killed by pesticides.
- Nitrates dissolved in soil water are absorbed into plants.
- Nitrates are used to produce amino acids which are synthesised into plant proteins.
- Animals consume plants or other animals to obtain amino acids for protein synthesis.
- Fertilisers can be added to soil to increase the nitrate content of the soil.
- Fertilisers can leach into fresh water, adding extra, unwanted nitrates. This will increase algal populations which can cause algal blooms. Algal blooms reduce light levels, killing aquatic plants. These dead plants, as well as dead algae, become food for bacteria which increase greatly in number. The bacteria use up large quantities of oxygen, reducing the oxygen availability for other organisms.
- Genetically modified (GM) crops are crops which have had their genetic material altered to enhance their characteristics.
- Genetically modified (GM) crops can be used to reduce the use of fertilisers.
- Pesticides sprayed onto crops can accumulate in the bodies of organisms over time. As they are passed along food chains, toxicity increases and can reach lethal levels.
- The build-up of toxic substances in living organisms is known as bioaccumulation.
- Biological control involves using a pest's natural predator to control pest numbers.
- Biological control and genetically modified (GM) crops can be used as alternatives to pesticides.

5.5 Extended response

Extended response

Q1:
Fertilisers added to crops can run off and add extra, unwanted nitrates to water systems. Describe the consequences of this process.

4 marks

5.6 Extension materials

DDT

DDT is a chemical which was developed as an insecticide and used during World War II to control diseases which were spread by insects such as malaria (spread by mosquitos). Although concerns about the use of DDT were raised from the beginning, it was a book written in 1962 by Rachel Carson called 'Silent Spring' which brought attention to the environmental impacts of DDT. In her book Carson documented the negative effects of pesticides on the environment, specifically the effects of DDT on birds. Carson described the bioaccumulation and biomagnification of DDT causing the death of birds, resulting in a "silent spring".

Silent spring

DDT is stored in the body fat of organisms and therefore bioaccumulates over time. It is also passed along food chains by biomagnification meaning top predators such as birds of prey were particularly vulnerable. Across the world, populations of birds of prey such as the peregrine falcon and osprey declined after the widespread use of DDT. Scientists have since discovered that DDT caused reproductive failure in these birds of prey as a result of egg shell thinning. Ultimately publication of 'Silent Spring' resulted in a ban on the use of DDT.

5.7 End of topic test

End of topic test: Food production — Go online

Farmers apply a mixture of chemicals to crops to increase crop yield.

Q2: One type of chemical added to the soil by farmers contains nitrates. Give the name of this type of chemical.

....

Q3: Why do plants require nitrates?

....

Q4: Another type of chemical added to the crop by farmers kills insects. Give the name of this type of chemical.

Aldrin is an insecticide which was used to control crop pests in the 1990s. Its use has since been banned due to its harmful effects on wildlife.

Q5: Aldrin is toxic to wildlife as it can build-up in their tissues causing damage. What term describes the build-up of toxic substances in living organisms?

....

Q6: Give one alternative to the use of insecticides such as aldrin.

Q7: Place the following statements into the correct order to show the consequences of excessive fertiliser use.

1. Extra nutrients cause algal populations to increase.
2. Dead plants, as well as dead algae, become food for bacteria which increase greatly in number.
3. Fertilisers leach into fresh water, adding extra, unwanted nitrates.
4. The bacteria use up large quantities of oxygen, reducing the oxygen availability for other organisms.
5. Overgrowth of algae reduce light levels, killing aquatic plants.

....

Q8: What term describes the overgrowth of algae?

Unit 3 Topic 6

Evolution of species

Contents

6.1	Mutation	258
6.2	Natural selection	259
6.3	Speciation	262
6.4	Learning points	265
6.5	Extended response	265
6.6	Extension materials	266
6.7	End of topic test	267

Learning objective

By the end of this topic you should be able to:

- define the term mutation;
- describe the possible effects of mutations on survival;
- state that mutations are spontaneous and are the only source of new alleles;
- name factors which can increase the rate of mutation;
- state that new alleles produced by mutation can result in plants and animals becoming better adapted to their environment;
- explain the importance of variation within a population;
- define the term adaptation;
- describe the process of natural selection;
- describe the process of speciation.

6.1 Mutation

A mutation is a random change to the genetic material of an organism. The change may occur at the level of the base sequence, in the example shown below one 'T' bases has been substituted for a 'C' base disrupting the base sequence of a gene.

Original sequence

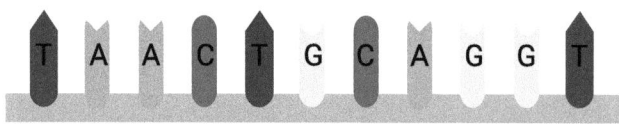

Point mutation

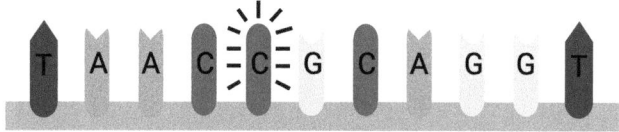

Mutation

Changes to the genetic material may involve changes to the structure of entire chromosomes, in the example shown below part of the chromosome has been lost meaning several genes have been lost.

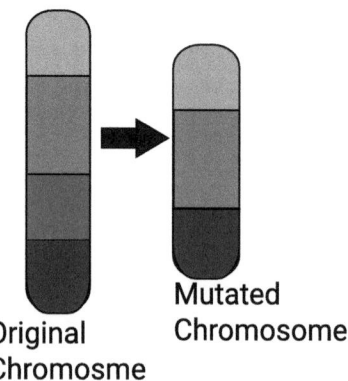

Original Chromosme   Mutated Chromosome

Mutation

Mutations are spontaneous meaning they mostly happen on their own and at random. Mutations are the only source of new alleles, giving rise to variation within a population.

A mutagen is the name given to anything which can increase the rate of mutation. Mutagens include:

- Radiation - X-rays, UV light, gamma rays
- Chemicals - colchicine, mustard gas and benzene

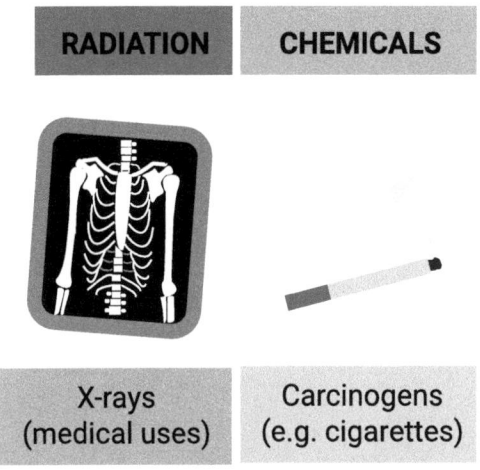

Mutagens

Mutations may be neutral, confer an advantage or a disadvantage to survival. Neutral mutations are changes in genetic material that are neither beneficial nor detrimental to the ability of an organism to survive. Some mutations confer an advantage to survival, for example the spread of lactose tolerance amongst the human population. Some mutations confer a disadvantage to survival, for example some mutations can cause cancer.

Mutations which confer an advantage to survival can result in plants and animals becoming better adapted to their environment. An adaptation is an inherited characteristic that makes an organism well suited to survival in its environment/niche. New alleles brought about by mutations result in variation which makes it possible for a population to evolve over time in response to changing environmental conditions.

6.2 Natural selection

Evolution is the process whereby all life on earth has developed from single celled organisms to multicellular organisms such as humans. Charles Darwin was the first person to propose the theory of evolution by natural selection.

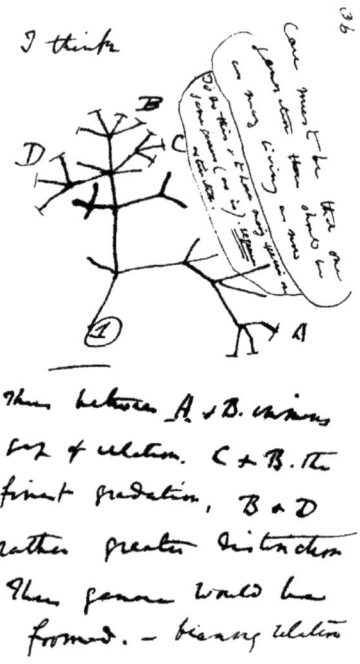

Charles Darwin's 1837 sketch of an evolutionary tree

Natural selection or survival of the fittest occurs when there are selection pressures. Selection pressures are any factors which can affect the survival or reproduction of an organism. For example availability of food, predation, climate or disease.

In natural selection or survival of the fittest, species produce more offspring than the environment can sustain. The best adapted individuals in a population survive to reproduce, passing on the favourable alleles that confer the selective advantage. These alleles then increase in frequency within the population. Over evolutionary time many small changes build up and alter the species. These small changes can result in a population becoming well adapted to its ecological niche and may eventually result in speciation.

Peppered moth evolution
The peppered moth is a nocturnal insect, which by day rests on tree trunks. It is preyed upon by insect eating birds such as robins. There are two forms, the speckled variety and the melanic variety (which arose by mutation).

Speckled variety

Melanic variety

In the early 19th century the speckled moths were well adapted to survive on **lichen** covered trees due to their camouflage. The melanic forms were less well adapted as they were easy for birds to see and catch. The best adapted individuals were the speckled moths because they were able to avoid predation. This means they were more likely to survive to reproduce and pass on the favourable speckled alleles which conferred a selective advantage to the next generation. These alleles increased in frequency within the population and so the number of speckled moths was high.

In the mid-19th century the industrial revolution took hold in Britain. This involved burning huge amounts of coal which produces sulphur dioxide and sooty smoke. Sulphur dioxide is lethal to some lichens and the soot settled on any exposed surfaces. The tree trunks turned black meaning the speckled moths were more easily seen and the melanic moths were better camouflaged. The best adapted individuals were now the melanic moths because they were able to avoid predation. This means they were more likely to survive to reproduce and pass on the favourable melanic alleles which conferred a selective advantage to the next generation. These alleles increased in frequency and so the number of melanic moths increased.

The change in the environment caused the numbers of melanic moths to increase and numbers of speckled moths to decrease in industrial areas. Moths are relatively short-lived and therefore this evolution by natural selection happened quite quickly.

Video: Natural selection in the rock pocket mouse   Go online

Watch this short video about the making of the fittest.

www.youtube.com/watch?v=wrTXvrKBlbc

6.3 Speciation

Speciation is the formation of new species in the course of evolution, it is brought about by mutations and natural selection.

A population of one species can only evolve into more than one species if groups within the population become isolated by barriers. Isolation barriers can be geographical, ecological or behavioural. Geographical barriers physically separate groups within the population, for example rivers or mountain ranges. Ecological barriers separate populations as a result of local differences in the environment, for example pH or salinity. Behavioural barriers separate populations as a result of the behaviour of individuals for example a group of females selecting for a different male characteristic during the mating season or populations adopting different **courting** patterns.

The process of speciation involves 4 main stages:

- Isolation by a barrier
- Mutation
- Natural selection
- Formation of separate species

Initially one population exists in the same area and interbreeds freely. The population then becomes split by a barrier (geographical, ecological or behavioural) into two sub-populations. Different mutations occur at random in each of the sub-populations causing new variation. Natural selection selects for different mutations in each group, due to different selection pressures. Each sub-population evolves until they become so genetically different that they are two different species. This means the two sub-populations can no longer interbreed and produce fertile offspring.

Speciation in the finches of the Galapagos

All the finches on the Galapagos islands have evolved from a common ancestor. The original population was split across several islands (geographical barrier) and experienced different mutations. Over time, as a result of natural selection, they have evolved different adaptations to their environments and are now separate species.

Stage 1:

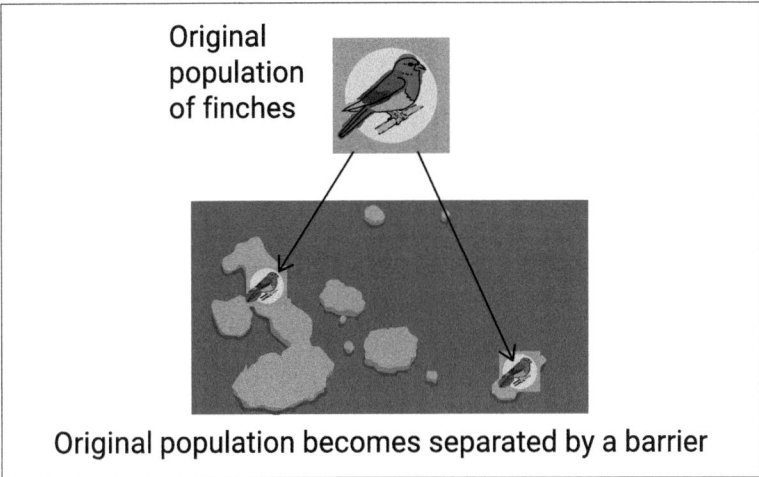

Stage 2:

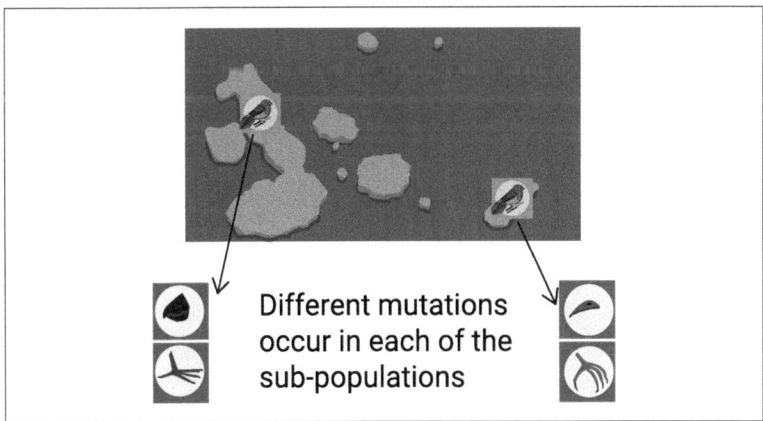

Stage 3:

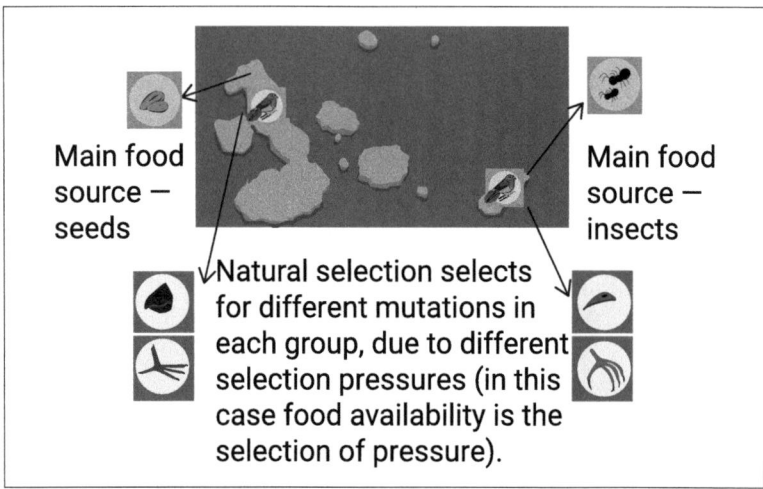

Stage 4:

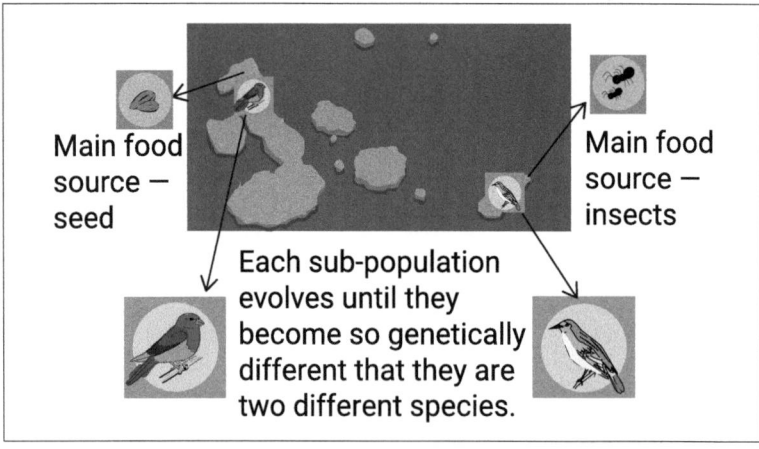

TOPIC 6. EVOLUTION OF SPECIES

6.4 Learning points

> **Summary**
>
> - A mutation is a random change to genetic material.
> - Mutations may be neutral, confer an advantage or a disadvantage to survival.
> - Mutations are spontaneous and are the only source of new alleles.
> - Environmental factors, such as radiation and some chemicals, can increase the rate of mutation.
> - New alleles produced by mutation can result in plants and animals becoming better adapted to their environment.
> - Variation within a population makes it possible for a population to evolve over time in response to changing environmental conditions.
> - An adaptation is an inherited characteristic that makes an organism well suited to survival in its environment/niche.
> - Species produce more offspring than the environment can sustain. Natural selection or survival of the fittest occurs when there are selection pressures.
> - Selection pressures are any factors which can affect the survival or reproduction of an organism.
> - The best adapted individuals in a population survive to reproduce, passing on the favourable alleles that confer the selective advantage. These alleles increase in frequency within the population.
> - Speciation occurs after part of a population becomes isolated by an isolation barrier, which can be geographical, ecological or behavioural.
> - Examples of each type of barrier, e.g. geographical - river, mountain range; ecological - pH, salinity; behavioural - a group of females selecting for a different male characteristic, populations adopting different courting patterns.
> - Different mutations occur in each sub-population. Natural selection selects for different mutations in each group, due to different selection pressures. Each sub-population evolves until they become so genetically different that they are two different species.

6.5 Extended response

Q1:
New species are formed by a process called speciation, describe this process.

4 marks

6.6 Extension materials

Charles Darwin

Charles Darwin was an English naturalist and biologist who authored a book called 'On the Origin of Species' where he described his theory of evolution.

Charles Darwin

Darwin proposed that animals and humans shared a common ancestor and had evolved as a result of natural selection. Darwin's theories were sparked while taking part in a five year voyage round the world on the HMS Beagle. Upon his return in 1836 he conducted detailed studies of the specimens he had collected on his travels. Through careful observations of plants and animals, Darwin noted that many species from all over the world were similar although each species showed adaptations specific to their local environment. This led him to believe that the species which exist today gradually evolved from common ancestors through a process called "natural selection".

> *"I have called this principle, by which each slight variation, if useful, is preserved, by the term of Natural Selection."*
> *- Quote from Charles Darwin*

Darwin argued that those organisms which were best adapted to their environment survived while those which were not best adapted failed to reproduce and therefore died off. Darwin's theory of evolution and the process of natural selection later became known simply as "Darwinism."

TOPIC 6. EVOLUTION OF SPECIES

6.7 End of topic test

End of topic test: Evolution of species Go online

Q2: Define the term mutation.

....

Q3: Name a factor which increases the rate of mutations.

....

Q4: Mutations are the only source of new alleles, giving rise to variation within the population. Explain why variation is important for the survival of a population.

The over use of antibiotics in society today has caused some types of bacteria to develop antibiotic resistance.

Q5: Place the following steps into the correct order to show the evolution of antibiotic resistance.

- Antibiotic resistance alleles increase in frequency within the population.
- Bacteria possessing the mutation survive as they have a selective advantage.
- In a large population of bacteria exposed to an antibiotic a random mutation occurs which confers resistance to the bacterium.
- Resistant bacteria pass on the favourable alleles (for antibiotic resistance) that confer the selective advantage.

....

Q6: Name the process described in the steps above.

....

Q7: Factors which can affect the survival or reproduction of an organism such as the presence of antibiotics in the example above are known as _____.

Q8: The stages below describe the process of speciation. Complete the missing stages

1. Speciation occurs after part of a population becomes isolated by an isolation barrier: _____

2. Natural selection selects for different mutations in each group, due to different selection pressures: _____

....

Q9: Match each example to the correct isolation barrier type.

List:

- river
- a group of females selecting for a different male characteristic
- mountain range
- populations adopting different courting patterns
- pH
- salinity

Behavioural	Geographical	Ecological

Unit 3 Topic 7
Life on Earth test

Life on Earth test

Go online

Ecosystems

The diagram below shows a South African food web.

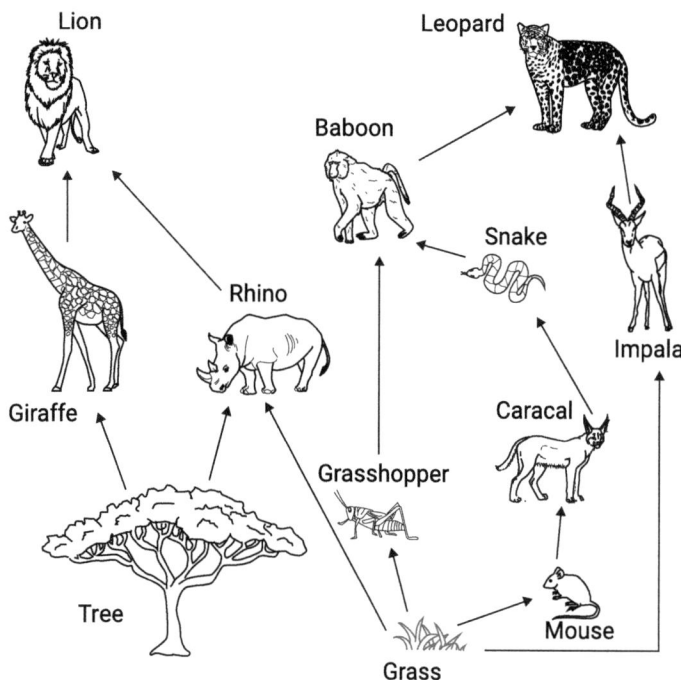

Q1: Name two organisms which are in competition with each other:

....

Q2: Complete the following sentence:

If the impala were killed due to over hunting the number of leopards would be likely to (increase/decrease/stay the same) due to _____.

....

Q3: Rhinos are a critically endangered species as a result of over-hunting. Define the term species.

....

Q4: If rhinos become extinct this will reduce the number of species in South Africa. What term describes the range of species in an ecosystem?

....

TOPIC 7. LIFE ON EARTH TEST

Q5: Two species of rhinos currently live in South Africa, both species eat grasses and other plants therefore they are in competition with each other. Name this type of competition.

....

Q6: Using information from the diagram give one ecological term to describe the lion.

Distribution of organisms

A group of students sampled the invertebrates living in a pond to determine pollution levels.

Q7: What name is given to organisms which by their presence or absence indicate environmental quality?

Biological keys can be used to identify organisms. The following table shows information about five different types of pond invertebrates.

Species	Antennae	Length (mm)	Legs
Stonefly nymph	present	8-12	3 pairs
Mayfly nymph	present	18-22	3 pairs
Water louse	present	8-12	6 pairs
Sludgeworm	absent	110-120	absent
Rat-tailed maggot	absent	50-60	absent

Q8: Use the information in the table above to complete the following paired statement key

1. Antennae present .... go to _____
   Antennae absent .... go to _____
2. _____ .... go to 4
   _____ .... Water louse
3. Length greater than 100mm .... Sludgeworm
   Length less than 100mm .... Rat-tailed maggot
4. Length greater than 15mm .... _____
   Length less than 15mm .... _____

....

Q9: The students measured factors in the pond such as pH. Complete the following sentence:

pH is (an abiotic/a biotic) factor.

....

Q10: Name the instrument used to measure pH and describe how it is used.

....

Q11: The students also investigated the biodiversity of invertebrates living in a nearby field. Name a sampling technique they could have used.

Photosynthesis

The diagram below represents the first stage of photosynthesis.

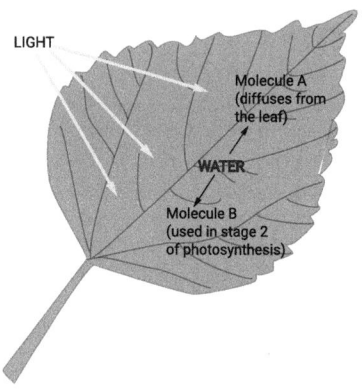

Q12: Name the first stage of photosynthesis and give the location where it takes place in a plant cell.

....

Q13: Name molecules A and B.

....

Q14: Identify one further product of stage 1 of photosynthesis which is required for stage 2.

....

Q15: Name stage 2 of photosynthesis.

....

Q16: Describe the reaction which takes place during stage 2 of photosynthesis.

An experiment was conducted to investigate the effect of light intensity on the rate of photosynthesis at different temperatures. The results are shown in the graph below.

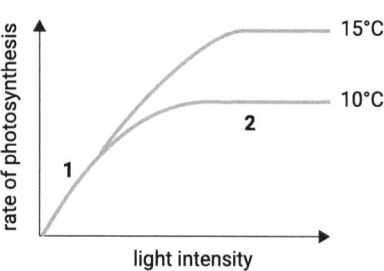

Q17: State the factor which is limiting the rate of photosynthesis at point 1 on the graph.

....

Q18: State the factor which is limiting the rate of photosynthesis at point 2 on the graph.

Energy in ecosystems

The following diagram shows a pyramid of numbers.

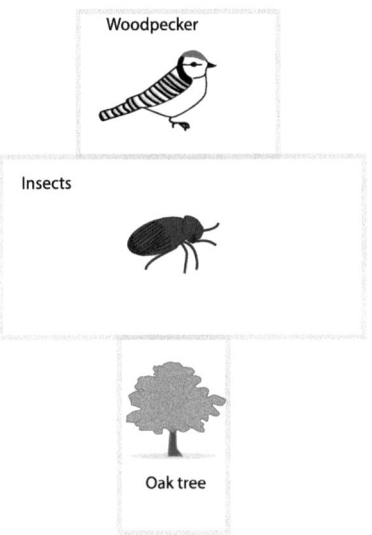

Q19: Give the meaning of the term pyramid of numbers.

....

Q20: Complete the following sentence:

Irregular shapes of pyramids of numbers based on different body sizes can be represented as true pyramids of _____.

....

Q21: Name one way energy is lost from a food chain.

Food production

Farmers use fertilisers containing nitrates to increase the yield of their crops.

Q22: Why do plants require nitrates?

....

Q23: Give one alternative to the use of chemical fertilisers.

....

Q24: Complete the missing stages to show the negative impacts of fertilisers on the environment.

1. Stage 1: Fertilisers can leach into fresh water, adding extra, unwanted nitrates.
2. Stage 2: _____
3. Stage 3: Algal blooms reduce light levels, killing aquatic plants.
4. Stage 4: _____
5. Stage 5: The bacteria use up large quantities of oxygen, reducing the oxygen availability for other organisms.

....

Q25: Pesticides applied to crops can negatively impact the environment by building up in the bodies of organisms over time. What term describes this build-up of toxic substances in living organisms?

....

Q26: Farmers may consider alternatives to pesticides to protect the environment. Give one alternative to the use of pesticides.

Evolution of species

Deer mice show variation in coat colour with some mice showing a dark coloured coat and others showing a light coloured coat.

The table below shows the results of an investigation into the numbers of the two varieties of mice found in two habitats.

Habitat	Light form	Dark form
Woodland with dark soils	11	67
Sandy hills	53	14

Q27: The light form of the mouse arouse due to mutation. Define the term mutation.

....

Q28: Name the process which has led to the results shown in the table.

....

Q29: Explain how this process has resulted in the high numbers of the light form of the mouse in the sandy hills.

....

Q30: The two varieties of the mice are still one species. To become different species an isolation barrier would need to separate the population. Name one type of isolation barrier.

Problem solving

An investigation was carried out to determine the effect of light intensity on the rate of photosynthesis.

Discs were cut from oak leaves and placed in syringes containing a solution that provided carbon dioxide. A procedure was used to remove air from the leaf discs to make them sink and the apparatus was placed in a darkened room.

The discs were then illuminated with a lamp at varying light intensities. As the leaves photosynthesised they floated. The time taken for three out of six discs to float was measured.

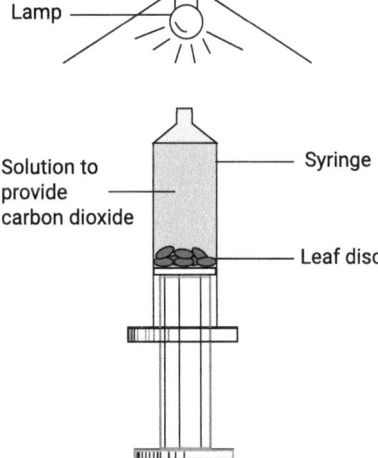

The results are shown in the table below.

Light intensity (kilolux)	Time taken for three discs to float (minutes)
10	25
20	18
30	11
40	7
50	3
60	3

Q31: Explain why it was good experimental procedure to use six discs at each light intensity.

....

Q32: An independent variable is the factor which scientists change throughout the course of an investigation. Identify the independent variable in this investigation.

....

Q33: Identify two factors which would need to be kept constant when setting up and conducting the experiment.

....

Q34: Calculate the percentage decrease in the time taken for three discs to float as the light intensity increased from 10 to 20 kilolux.

....

Q35: Plot a line graph to display the results shown in the table.

....

Q36: Describe the relationship between light intensity and the time taken for three discs to float.

....

Q37: Predict the time taken for three discs to float if the experiment was carried out at a light intensity of 45 kilolux.

Problem solving

Chemical fertiliser versus organic fertiliser

Scientists have a theory that chemical fertiliser affects plant growth differently compared to organic fertiliser.

In a study, lettuce plants were exposed to different treatments. Five lettuce plants were grown in soil containing no additional fertiliser, five lettuce plants were grown in soil containing chemical fertiliser and five lettuce plants were grown in soil containing organic fertiliser.

After 50 days the scientists measured the shoot height of each plant. They found that the average height of the plants grown in soil containing no additional fertiliser was 20cm, the average height of the plants grown in soil containing chemical fertiliser was 27cm and the average height of the plants grown in soil containing organic fertiliser was 24cm.

The differences may seem small, however, across an entire crop it could mean that farmers are able to grow larger lettuce plants more quickly and therefore increase their yield significantly.

Q38: Suggest the aim of the research described in the passage.

....

Q39: A dependent variable is what scientists measure or observe as a result of the changes they make in their investigation. Identify the dependent variable in this investigation.

....

Q40: What term describes the group of plants which were grown in soil containing no additional fertiliser?

....

Q41: Complete the table to show the results of the study.

Treatment	Plant height (cm)

....

Q42: What conclusion did the scientists draw from this study?

....

Q43: Give a reason why it could be suggested that the results of the investigation might be unreliable.

Appendix

A Apparatus and techniques	281
A.1 Apparatus	282
B Laboratory techniques	287
C Field techniques	293

Unit 4 Appendix A

Apparatus and techniques

Contents
A.1 Apparatus 282

> **Learning objective**
>
> By the end of this topic you should be able to:
>
> - describe the uses of the following pieces of apparatus:
>   - beaker;
>   - balance;
>   - measuring cylinder;
>   - dropper / pipette;
>   - test tube / boiling tube;
>   - thermometer;
>   - funnel;
>   - syringe;
>   - timer / stopwatch;
>   - microscope;
>   - petri dish;
>   - quadrat;
>   - pitfall trap;
>   - light / moisture meter;
>   - water bath;
> - describe the following techniques:
>   - measuring enzyme activity;
>   - using a respirometer;
>   - measuring transpiration using a potometer;
>   - measuring the rate of photosynthesis;
>   - measuring abiotic factors;
>   - measuring the distribution of a species;
>   - using a transect line.

A.1 Apparatus

Apparatus	Diagram	Function
beaker		A glass container used to hold liquids.
balance		A piece of equipment used to measure mass.
measuring cylinder		A graduated cylinder used to measure the volume of liquids.

APPENDIX A. APPARATUS AND TECHNIQUES 283

dropper/pipette	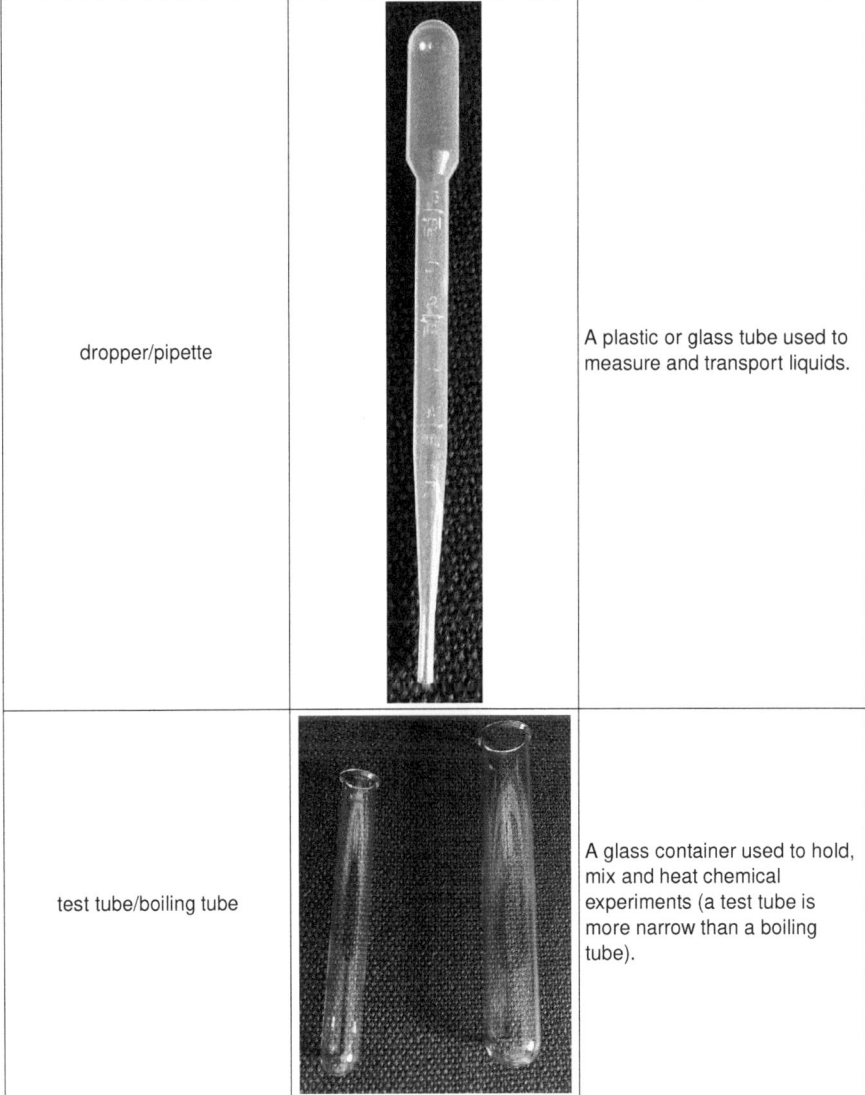	A plastic or glass tube used to measure and transport liquids.
test tube/boiling tube		A glass container used to hold, mix and heat chemical experiments (a test tube is more narrow than a boiling tube).

© HERIOT-WATT UNIVERSITY

thermometer		A device used to measure temperature.
funnel		A cone shaped piece of plastic or glass which allows liquids to be transferred from one container to another.

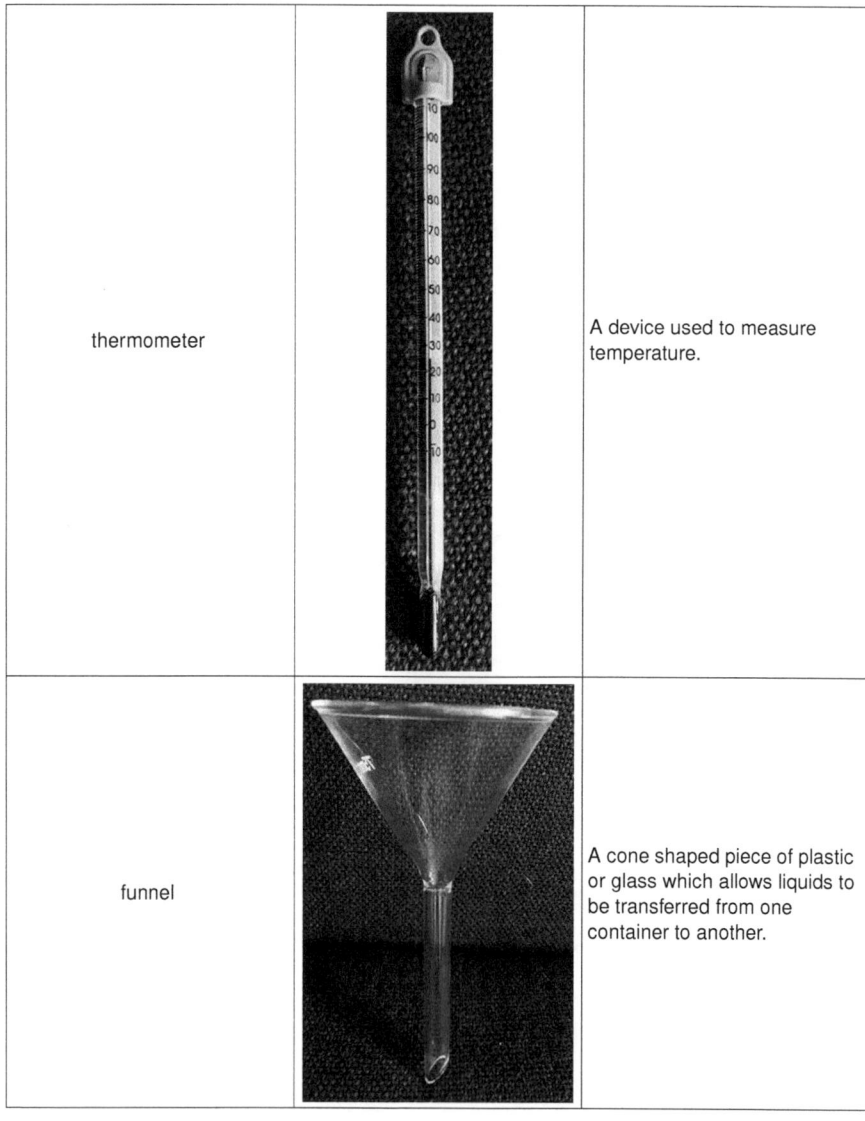

APPENDIX A. APPARATUS AND TECHNIQUES

syringe		A plastic tube with a plunger used to measure and transport liquids.
timer/stopwatch		A device used to measure time.
microscope		A device which magnifies a specimen using a series of lenses.

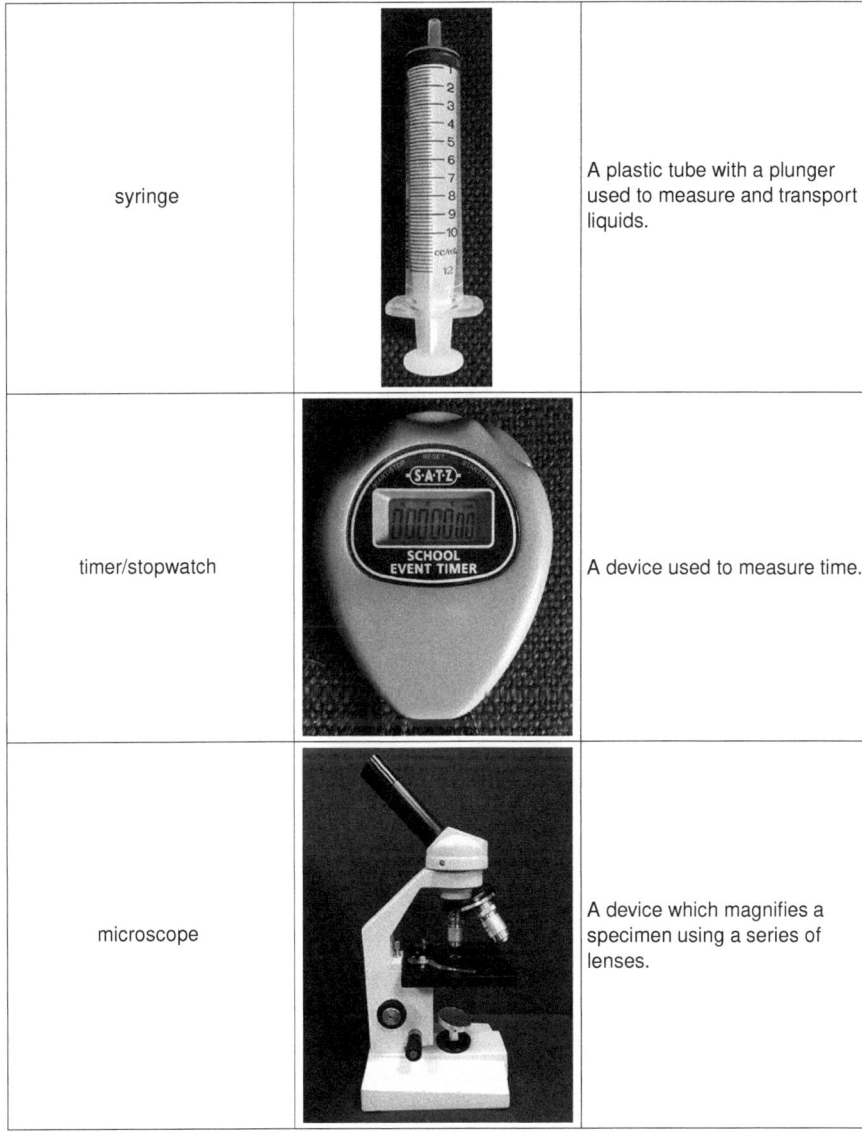

petri dish		A circular plastic or glass dish often used in microbiology with nutrient agar to grow cells.
quadrat		A square grid used to sample plants or slow moving animals (e.g. limpets).
pitfall trap		A container which is sunk into the ground to sample invertebrates.
light meter / moisture meter		A device used to measure light intensity. / A device used to measure moisture.
water bath		A water filled container which maintains a constant water temperature.

Unit 4 Appendix B

Laboratory techniques

Measuring enzyme activity
Experiments can be conducted to measure the activity of an enzyme. The experiment conducted will depend on the type of enzyme. The experiment shown below can be used to measure the activity of the enzyme catalase.

The enzyme catalase breaks down hydrogen peroxide into water and oxygen. The following experiment uses the release of oxygen to determine the effect of temperature on enzyme activity. An enzyme extract can be produced from a tissue which contains catalase such as potato. Test tubes of potato catalase extract and hydrogen peroxide can be placed in waterbaths at varying temperatures. Discs of filter paper can be dipped in the enzyme extract and placed at the bottom of a test tube of hydrogen peroxide. As the reaction occurs oxygen is released and the paper disc will begin to float. The time taken for the paper disc to float to the surface of the hydrogen peroxide can be measured. The more quickly the disc floats the more active the enzyme.

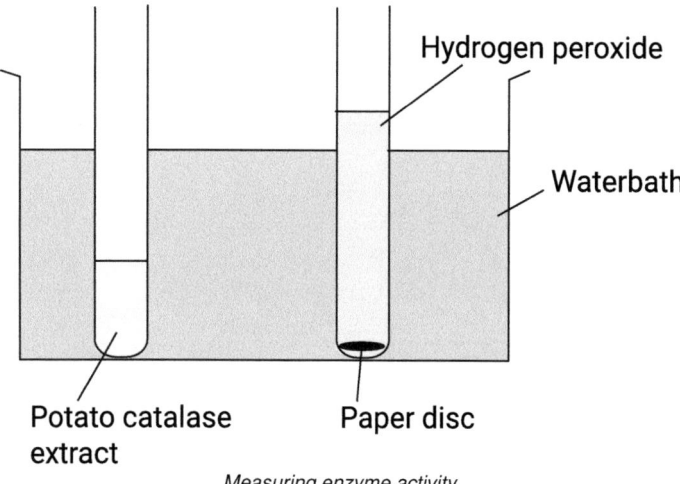

Measuring enzyme activity

Using a respirometer

A respirometer is a set up used to measure the rate of respiration in a small organism. A respirometer measures oxygen consumption. An organism is placed into the respirometer and the chamber is sealed. The organism begins to respire and uses up the oxygen in the sealed apparatus, a chemical is included in the tube which absorbs the carbon dioxide released by an organism. As the oxygen is used up the pressure in the chamber decreases and the coloured dye in the manometer moves along the tube towards the respiring organism. The syringe can be used to return the manometer level to its starting point and measure the volume of oxygen consumed by the organism.

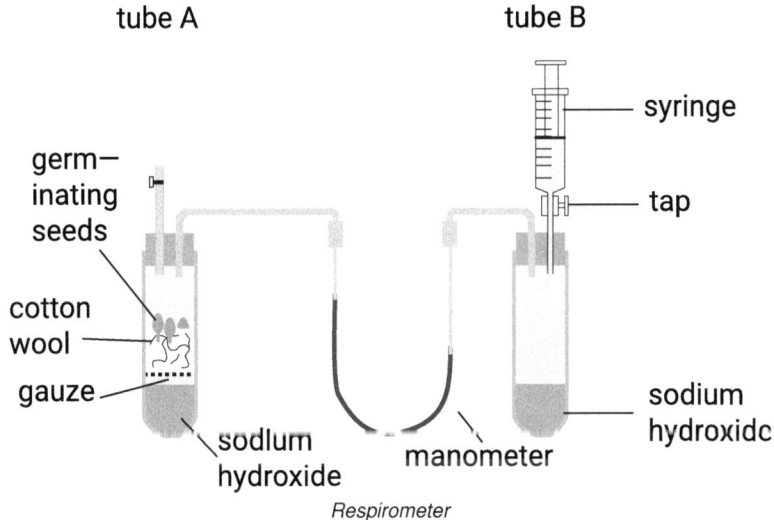

Respirometer

Measuring transpiration using a potometer
A potometer measures the rate of transpiration in a leafy shoot by measuring the water uptake of the shoot.

A bubble potometer is shown in the diagram below. As the experiment runs the cut shoot takes up water and the rate of transpiration is determined by measuring the distance travelled by the air bubble over a set period of time. The greater the distance travelled by the bubble the greater the rate of transpiration. The apparatus can be set up to investigate the effect of several different factors on the rate of transpiration for example the addition of a fan can be used to investigate the effect of wind speed on transpiration.

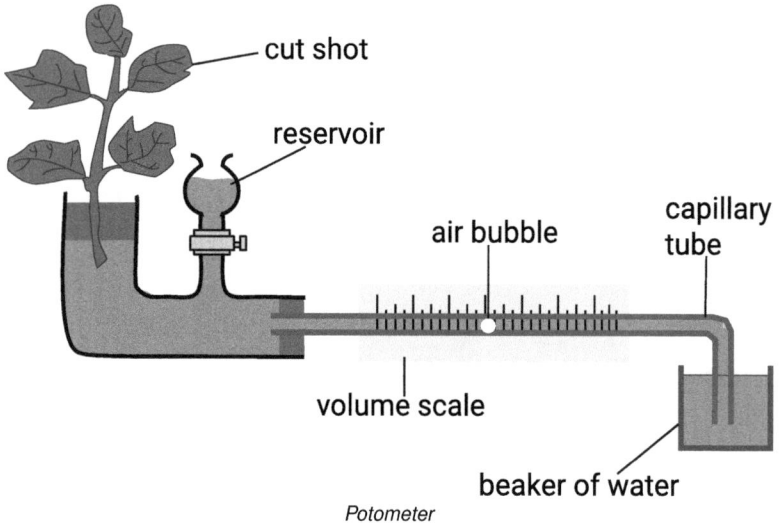

Potometer

Measuring the rate of photosynthesis

Experiments can be conducted to measure the rate of photosynthesis. These experiments measure the uptake of carbon dioxide, production of oxygen/carbohydrate or increase in dry mass to determine the rate of photosynthesis.

The experiment below shows a method of measuring the production of oxygen using an aquatic plant such as Elodea. The plant is placed in a solution which provides it with carbon dioxide and an inverted water filled measuring cylinder is placed above it to capture and measure the oxygen the plant produces. Factors such as carbon dioxide concentration and light intensity can be altered to determine their effect on the rate of photosynthesis.

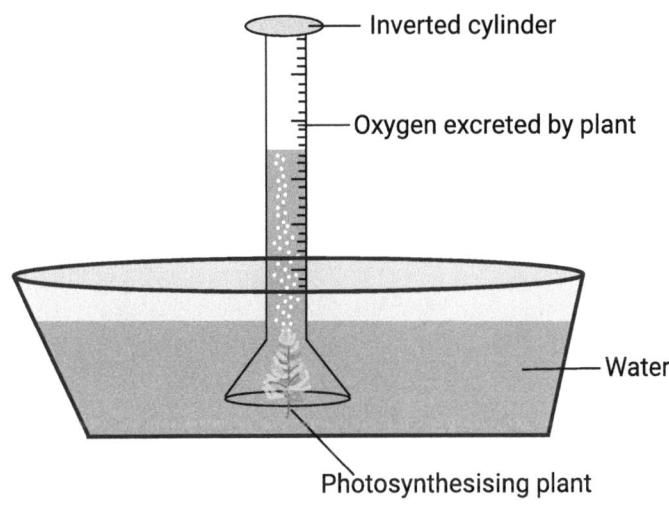

Measuring the rate of photosynthesis

Unit 4 Appendix C

Field techniques

Measuring abiotic factors

Light intensity is measured using a light meter. The sensor is held upward and a reading is taken from the scale on the meter. Soil moisture is measured using a moisture meter. The probe is pushed into the ground and a reading is taken from the scale on the meter. A pH meter is used to measure pH. This meter is used in the same way as a moisture meter.

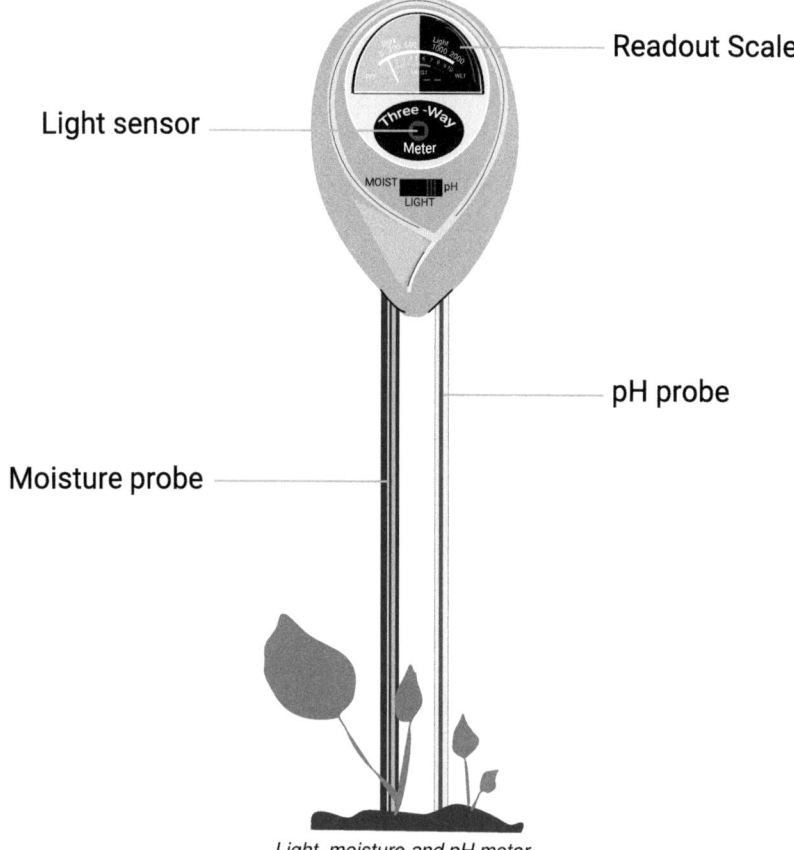

Light, moisture and pH meter

Temperature is measured using a thermometer. The thermometer is placed in the experimental area, allowed to stabilise and a reading is taken from the scale.

APPENDIX C. FIELD TECHNIQUES 295

Thermometer

Measuring the distribution of a species
In order to measure the distribution of a species throughout a habitat, samples of the species under investigation must be collected for example through the use of pit fall traps.

Pitfall traps are used to sample invertebrates which live on the ground. A hole is dug into the ground and a plastic pot is placed inside. It is important that the trap is level with the ground to ensure the invertebrates can fall in. The trap is camouflaged and left for a set period of time after which the number of each invertebrate found within the trap is counted.

© HERIOT-WATT UNIVERSITY

Pitfall trap

When sampling invertebrates using pitfall traps it is important to set several traps across the investigated area and calculate averages to ensure the results are representative of the area being investigated.

Where it is not possible to collect organisms, their distribution can be determined by sampling. Quadrats are used for estimating the abundance of plants or slow moving animals (such as limpets) in an ecosystem. The quadrat is thrown at random and the number of squares that contain the investigated organism are counted. To be representative of the total population an adequate number of quadrats are randomly thrown. An average number of each organism per quadrat is calculated then scaled up represent the area.

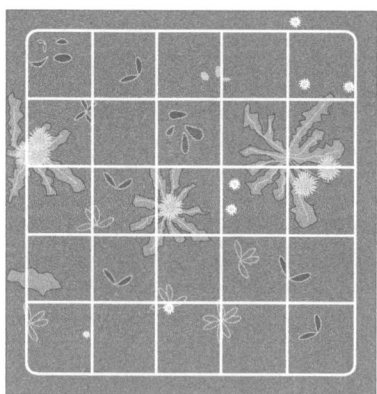

Quadrat

Quadrats only sample a small percentage of the ecosystem as a whole therefore to ensure the results are representative the quadrat should be thrown randomly many times and an average calculated.

Using a transect line
A transect line is a tape or rope laid along the ground across a habitat, organisms can be sampled along the transect line to determine their distribution.

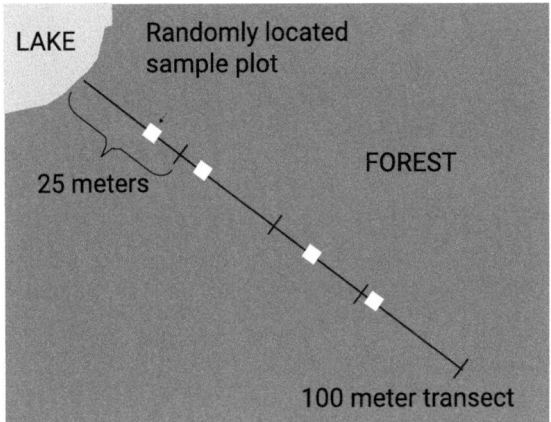

Line transect

Glossary

Aerobic
> requiring oxygen

Allele
> The form of a gene.

Alveoli
> Tiny sacs for gas exchange in lungs.

Anther
> Structure in a flower that produces pollen grains.

Aorta
> The main artery which carries blood away from the heart.

Artery
> Blood vessels that carry oxygenated blood away from the heart.

Atria
> Upper chambers of the heart that pass blood to the lower ventricles.

Bilayer
> a structure composed of two layers of molecules

Biomass
> the total quantity of organisms in a given area.

Brain
> Organ of the central nervous system that controls vital functions.

Capillaries
> Tiny blood vessels that are only one cell thick for efficient material exchange.

Catalyst
> a substance which speeds up a chemical reaction but remains unchanged at the end of the process.

Cellulose
> a carbohydrate which is a component of the cell wall in plants

Central Nervous System (CNS)
> Part of the nervous system made up of the brain and spinal chord.

Cerebellum
> Section of the brain that controls coordination, movements and balance.

Cerebrum
> Section of the brain that controls memory, conscious thoughts, intelligence and emotions.

GLOSSARY

Chromatid
Replicated copy of a chromosome

Chromosome
Composed of DNA. Codes for all of an organisms characteristics

Chromosome compliment
The number of chromosomes found in a cell

Complementary
matching in shape/structure

Continuous
Variations that seen as one extreme to the other with a range of values in between.

Courting
a collection of behaviours which are used to attract a mate.

Crustacean
a type of invertebrate with a hard exoskeleton, this group includes crabs, krill, lobsters and crayfish.

Diploid
A cell that contains a double set of chromosomes

Discrete
Variations that are able to be categories into groups.

Distribution
the way in which something is spread across an area.

Dominant
The form of a gene which is always expressed.

Endocrine gland
Gland that secretes hormones into the bloodstream.

Equator
Middle position of a cell where chromosomes align and attach to spindle fibres in mitosis

Exponentially
more and more rapidly.

Fertile
capable of producing offspring.

Fertilisation
The fusion of the nuclei of two gametes.

Gamete
Sex cells.

Genotype
   The particular alleles that an organisms has for a genotype.

Glucagon
   Hormone produced by the pancreas which triggers glycogen conversion into glucose.

Guard cells
   Control the opening and closing of stoma.

Haemoglobin
   Component of red blood cells that aids the transport of oxygen.

Heterozygous
   Two different alleles of a genotype ie. Aa or Bb.

Homozygous
   Two alleles the same for a genotype ie. AA or aa.

Insulin
   Hormone produced by the pancreas which triggers glucose conversion into glycogen.

Inter neuron
   Nerve cell that are found in the CNS where they connect with other neurons.

Lacteal
   Vessel in the villi that is responsible for transporting fats.

Lichen
   a symbiotic association between a fungus and algae.

Lignin
   Lining of the xylem made of carbohydrate that provides strength and support.

Liver
   Large organ involved in blood glucose control.

Lymph
   Liquid that transports the products of fat digestion from the lacteal.

Medulla
   Section of the brain that controls breathing and heart rate.

Mitosis
   A process of cell division that produces two genetically identical daughter cells

Motor neuron
   Nerve cell that carries electrical impulses from the CNS to muscles and glands (effectors).

Organ
   A group of similar tissues that work together to carry out the specific function

Organelle
  a membrane bound structure which performs a specific function within the cell.

Ovary
  Female sex organ.

Ovule
  Structure containing female gametes in plants.

Pancreas
  Organ that produces digestive enzymes and the hormones glucagon and insulin.

Parasitoid
  an insect whose larvae live as parasites inside a host, the larvae feed from the host and eventually kill them.

Peristalsis
  Waves of muscular contractions that help food move through the alimentary canal.

Pharmaceutical
  relating to medicinal drugs

Phenotype
  The physical appearance expressed by an organisms due to their genotype.

Phloem
  Vessel in plants that transports sugars.

Phospholipid
  a lipid found in the cell membrane

Pollen grain
  Structure containing male gametes in plants.

Pollination
  The transfer of pollen from the anthers to a stigma in plants.

Polygenic
  Type of inheritance involving several genes acting together.

Pulmonary artery
  Artery that carries deoxygenated blood from the heart to the lungs.

Pulmonary vein
  Vein that carries oxygenated blood from the lungs to the heart.

Recessive
  The form of a gene which will only be expressed if the genotype is homozygous

Reflex arc
  Pathway of information from a sensory neuron to a motor neuron via the interneurons.

Replication
    The copying of a cell's DNA before mitosis

Sensory neuron
    Nerve cell that carries electrical impulses from sense organs to CNS.

Soluble
    able to be dissolved

Specialised cell
    A cell that has become differentiated to perform a specific function

Sperm cell
    Gamete produced in the testes of male animals.

Spindle fibres
    Protein threads that pull chromatids apart during mitosis

Spongey mesophyll
    Loosely packed plant leaf tissue with air spaces for gas exchange.

Stem cell
    Unspecialised animal cell that is capable of dividing into cells that could become many different cell types

Stomata
    Tiny pores that allow for gas exchange in the leaf epidermis.

Synapse
    Space/gap between two neurons.

System
    A group of organs that work together to carry out the specific function

Target organ
    Organ with receptors that recognises specific hormones.

Temperate
    a region characterised by mild temperatures.

Testes
    Male sex organ.

Tissue
    A group of similar cells that carry out the same function

Transpiration
    The loss of water by evaporation through the stomata of leaves.

Variation
    Differences in the characteristics of members of the same species.

GLOSSARY

Vein
Blood vessels that carry blood back to the heart.

Vena cava
Blood vessels that carry deoxygenated blood to the heart from the body.

Ventricles
Lower chambers of the heart that receive blood from the upper arteria

Villi
Finger-like projections in the small intestine that provide a large surface area for absorbing food.

Xylem
Narrow, dead tubes with lignin walls that transports water and minerals in plants.

Zygote
The fertilised egg.

Hints for activities for Unit 2

Topic 2: Control and communication

Practical: Reflex actions and the reflex arc

Hint 1:

Help with the method:

Use the following basic information to refine your own method/ procedure or to help you get started.

1. Subject 1 should hold out the chosen hand and extend the thumb and index finger so they are 8 cm apart.
2. The partner should hold a ruler with the '0cm' end in line with the subject 1's extended thumb and index finger. The ruler should be vertical.
3. The ruler is dropped, and the subject grasps it between the thumb and index finger.
4. Record the number at the subject's fingertips, i.e. distance the ruler fell through the subject's fingers.

Answers to questions and activities for Unit 1

Topic 1: Cell structure

Ultrastructure of an animal cell (page 10)

Q1:

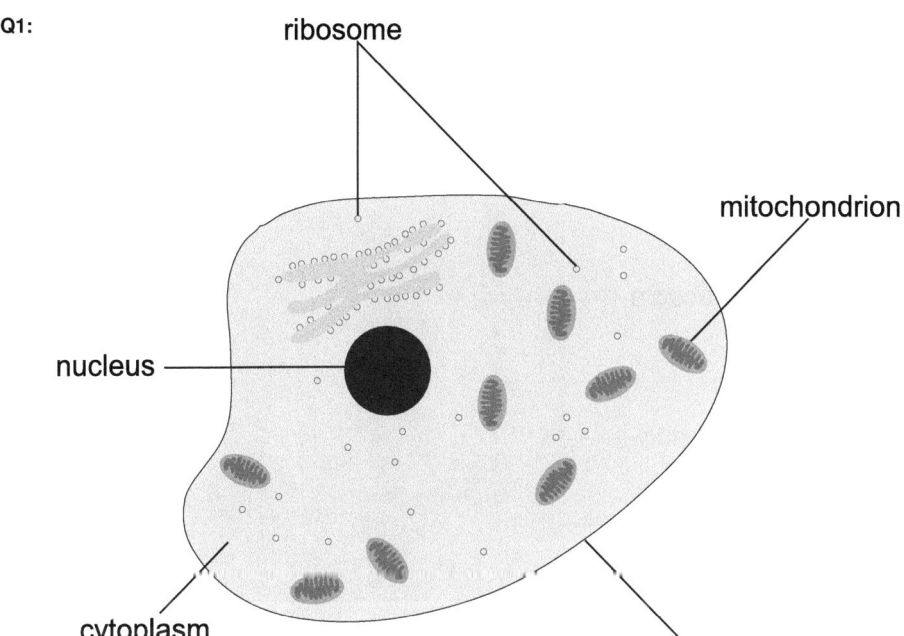

Ultrastructure of a plant cell (page 11)

Q2:

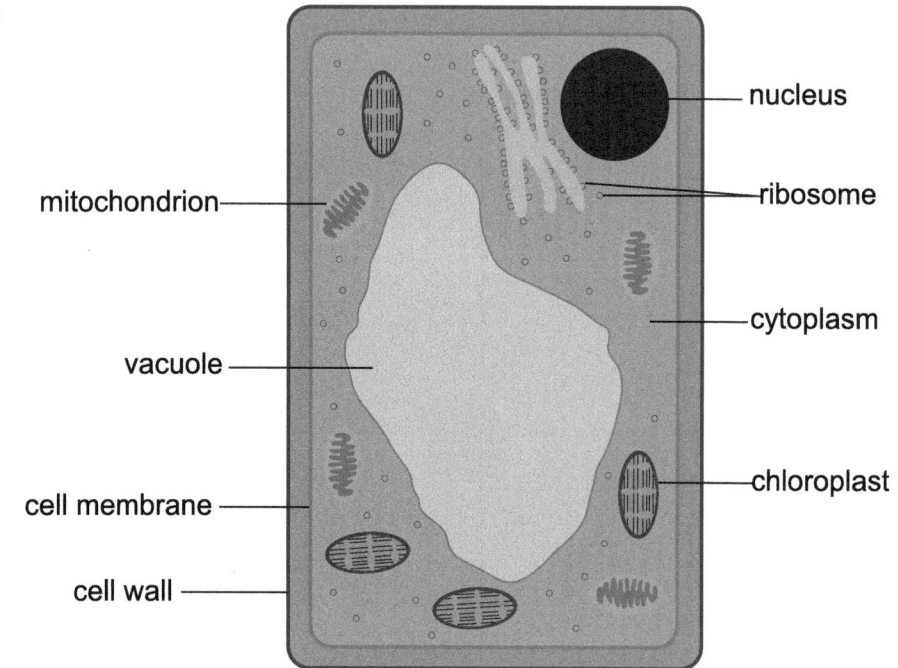

Ultrastructure of a fungal cell (page 12)

Q3:

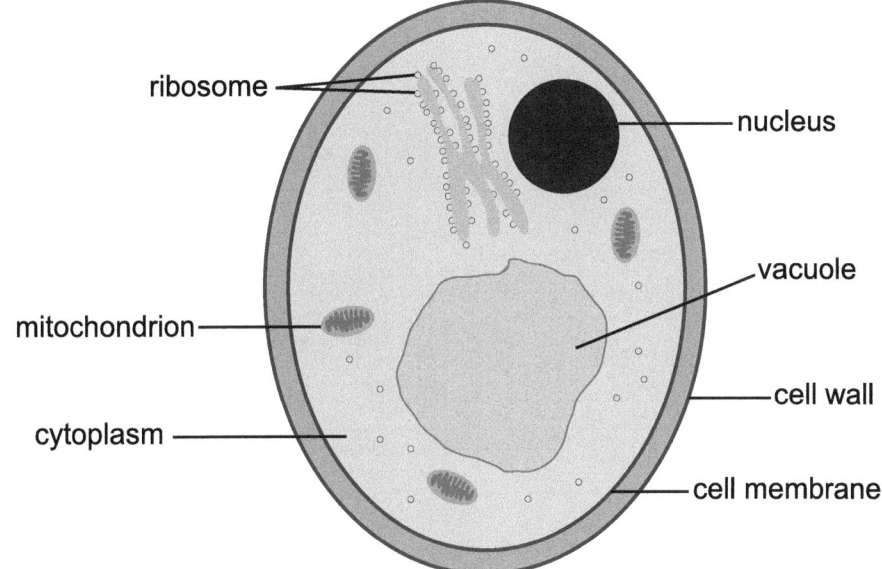

Ultrastructure of a bacterial cell (page 13)

Q4:

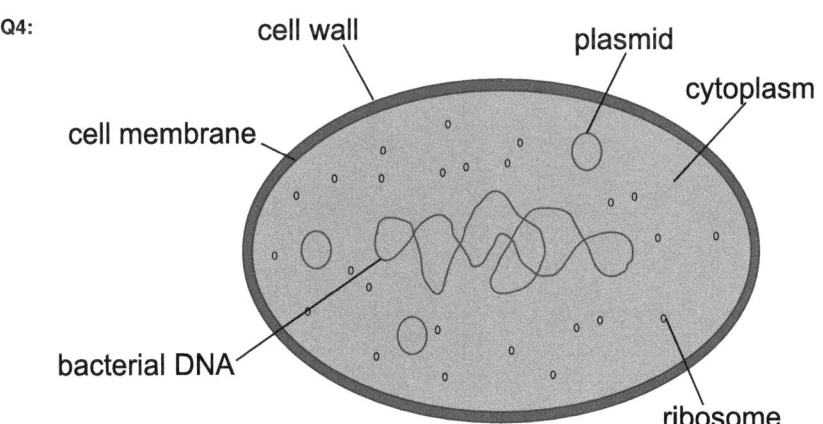

Structure and function of a cell (page 14)

Q5:

Structure	Function
Nucleus	Controls cell activities
Cell membrane	Controls entry and exit of molecules
Cytoplasm	Site of chemical reactions
Ribosome	Site of protein synthesis
Mitochondrion	Site of aerobic respiration
Chloroplast	Site of photosynthesis
Vacuole	Stores water, sugar and salts in a solution called cell sap
Cell wall	Gives the cell a rigid structure
Plasmid	Contains additional genes which are beneficial to the cell

Cell structures (page 14)

Q6:

Animal cell	Plant cell	Fungal cell	Bacterial cell
nucleus	nucleus	nucleus	cell membrane
cell membrane	cell membrane	cell membrane	cytoplasm
cytoplasm	cytoplasm	cytoplasm	ribosomes
ribosomes	ribosomes	ribosomes	cell wall
mitochondria	mitochondria	mitochondria	plasmids
	chloroplasts	vacuole	
	vacuole	cell wall	
	cell wall		

End of topic test: Cell structure (page 17)

Q7: Animal, Plant, Fungal

Q8: Plant

Q9: Plant, Fungal, Bacterial

Q10: Controls cell activity/activities

Q11: Cell membrane

Q12: Ribosome

Q13:

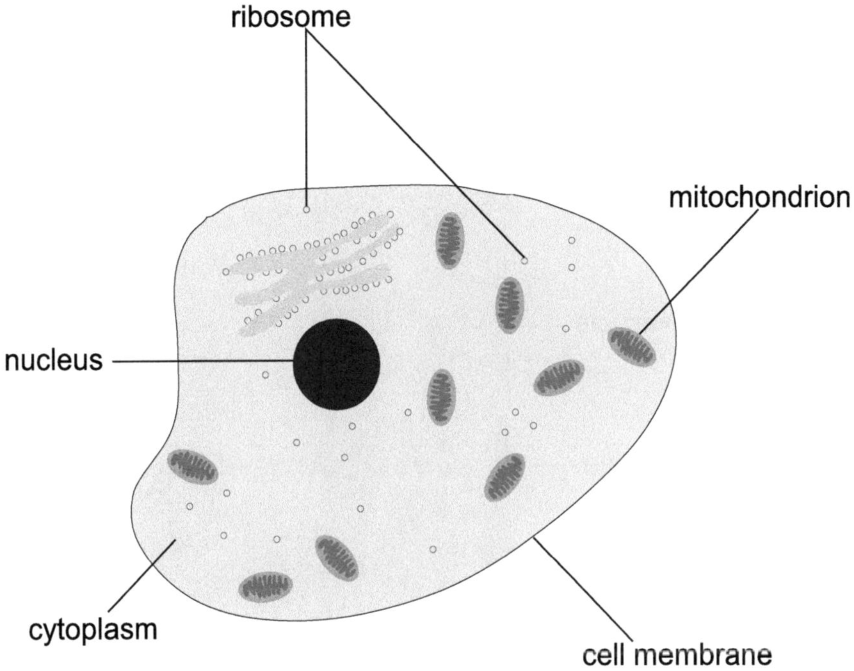

- ribosome
- mitochondrion
- nucleus
- cytoplasm
- cell membrane

Q14:

nucleus
ribosome
mitochondrion
cytoplasm
vacuole
chloroplast
cell membrane
cell wall

Q15:

ribosome
nucleus
vacuole
mitochondrion
cell wall
cytoplasm
cell membrane

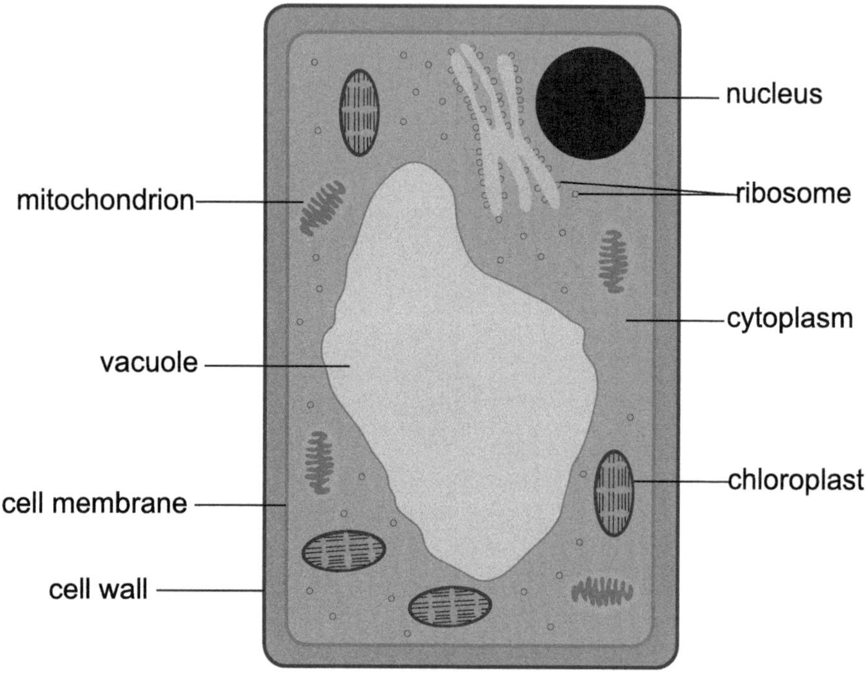

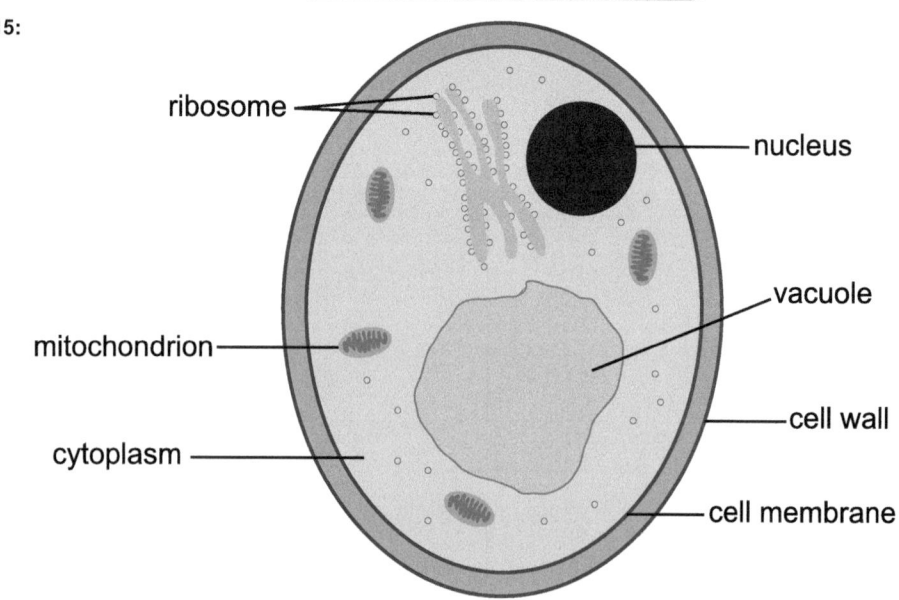

Q16:

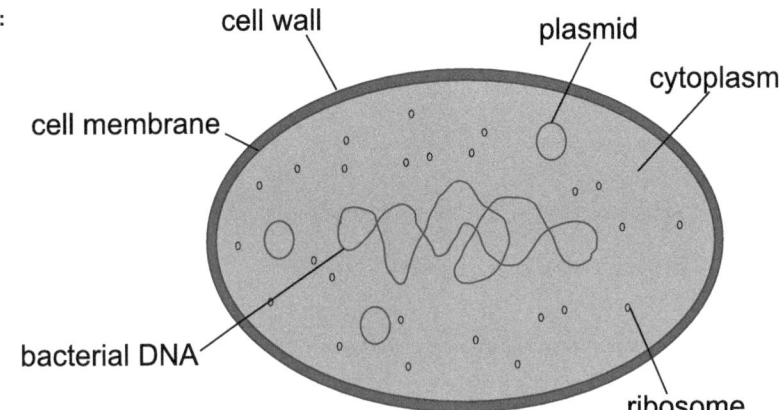

Topic 2: Transport across cell membranes

Extended response question (page 27)

Q1: Any 3 from:

- Water moves by osmosis.
- Water moves in (to the model cell).
- From higher water concentration to lower water concentration/down the water concentration gradient.
- Through a selectively permeable membrane.

End of topic test: Transport across cell membranes (page 28)

Q2:

X phospholipid, Y protein

Q3: Protein

Q4: Osmosis

Q5:

Water moved from a **higher** water concentration **inside** the red blood cells to a **lower** water concentration **outside** the red blood cells.

Q6: plasmolysed

Q7:

Passive transport	Active transport
Does not require energy	Requires energy
Moves molecules from a higher concentration to a lower concentration	Moves molecules from a lower concentration to a higher concentration
Moves molecules down a concentration gradient	Moves molecules against a concentration gradient

Q8: Osmosis and diffusion

Q9: Oxygen OR glucose

Q10: Carbon dioxide

Topic 3: DNA and the production of proteins

DNA (page 34)

Q1:

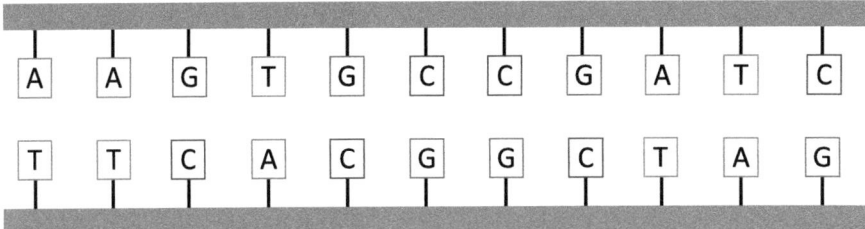

Extended response questions (page 35)

Q2:

1. DNA takes the form of a double-stranded helix.
2. The DNA bases are adenine, cytosine, guanine and thymine.
3. Adenine/A is always paired with thymine/T and cytosine/C is always paired with guanine/G.

Q3: Any four points from:

1. DNA carries the genetic information for making proteins.
2. DNA bases/adenine, cytosine, guanine and thymine make up the genetic code.
3. The base sequence determines amino acid sequence in proteins.
4. A gene is a section of DNA which codes for a protein.
5. mRNA carries a complementary copy of the genetic code from the DNA, in the nucleus, to a ribosome.
6. Proteins are assembled from amino acids.

End of topic test: DNA and the production of proteins (page 37)

Q4:

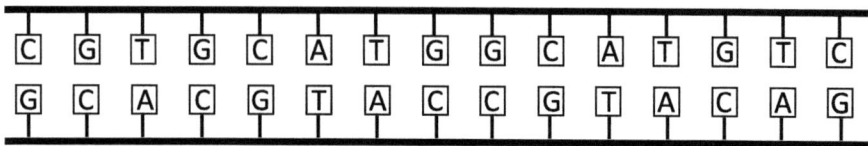

Q5:

- A = adenine
- T = thymine
- C = cytosine
- G = guanine

Q6: Sequence of bases/order of bases/base order/base sequence

Q7: Protein OR proteins

Q8: mRNA

Topic 4: Proteins

Extended response question (page 47)

Q1:

1. Any correct named enzyme and substrate for example phosphorylase and glucose-1-phosphate.
2. Substrate/glucose-1-phosphate **built-up** into larger molecules.

For marks 3. and 4. any two from:

- Enzyme speeds up chemical reactions.
- Active site is specific/complementary fit/enzyme-substrate complex forms/substrate binds to the active site.
- Named products (i.e. if using the example phosphorylase example above the product is starch).
- Product released from enzyme/active site.

End of topic test: Proteins (page 49)

Q2:

Type of protein	Function
Structural	offer support to the cell/organism
Receptor	allow signals to be transmitted across the membrane into the cell
Enzyme	speed up cellular reactions and are unchanged in the process
Antibody	combine with pathogens to destroy them and protect the body from disease
Hormone	act as chemical messengers carrying information from one part of the organism to another

Q3:

The proteases within the biological washing powders are carrying out a **degradation** reaction.

In this reaction the proteins are the **substrate** and the peptides are the **product** of the protease enzymes.

Q4:

Prediction: decrease

Explanation: Enzyme/protease denatured OR enzyme/active site has changed shape

Q5: pH

Q6:

Enzymes **speed up** chemical reactions, they allow reactions to occur at **lower** temperatures.

Enzymes are **unchanged** by the chemical reaction they catalyse.

Q7: 2,1,3

Q8: Active site

Q9: Joining/building up molecules OR simple/small to complex/large molecules

Q10: Protein/amino acids

Topic 5: Genetic engineering

Extended response question (page 54)

Q1: Any five from:

1. Required gene identified/located.
2. Required gene extracted.
3. Plasmid removed from bacterial cell.
4. Gene inserted into plasmid.
5. Plasmid inserted into (new) bacterial/host cell.
6. Bacterial cells grown/cultured/multiply.
7. Required product extracted/purified/made.

End of topic test: Genetic engineering (page 56)

Q2:

Stage 1: identify section of DNA that contains required gene from source chromosome
Stage 2: extract required gene
Stage 3: extract plasmid from bacterial cell
Stage 4: insert required gene into bacterial plasmid
Stage 5: insert plasmid into host bacterial cell to produce a genetically modified (GM) organism

Q3: enzymes

Q4: plasmid

Q5: insert plasmid into host bacterial cell.

Topic 6: Respiration

Extended response question (page 65)

Q1: Any four from:

1. Occurs in the absence of oxygen.
2. Glucose is broken down to (two molecules of) pyruvate.
3. Two molecules of ATP produced.
4. Pyruvate converted into lactate.
5. Occurs in the cytoplasm.

End of topic test: Respiration (page 65)

Q2: Aerobic respiration

Q3: Any of the following:

- muscle contraction
- cell division
- movement
- mitosis
- protein synthesis
- transmission of nerve impulses
- active transport

Q4: pyruvate

Q5: 2

Q6: enzymes

Q7: mitochondria

Q8: fermentation

Q9: lactate

Q10: cytoplasm

Q11: B C D

Q12: A B

Q13: mitochondria OR mitochondrion

Topic 7: Cell biology test
Cell biology test (page 68)

Q1: A = cytoplasm, B = vacuole

Q2: Control cell activities

Q3: ribosome

Q4: It would have chloroplasts.

Q5: cellulose

Q6: C

Q7: turgid

Q8: plasmolysed

Q9:
Animal cells do not have cell walls. (1 mark)
Cell wall prevents the cell from bursting (1 mark)

Q10: A = phospholipid, B = protein

Q11: double helix or helix

Q12: mRNA

Q13: ribosome

Q14:
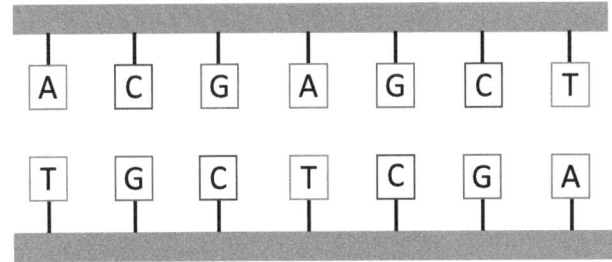

Q15: Speed up reactions and remain unchanged by the process.

Q16: Active site

Q17: degradation

Q18: It would become denatured/it would change shape/it would no longer function.

Q19: pH or temperature

Q20: optimum

Q21: enzyme

Q22: plasmid

Q23: insert plasmid/structure Y into host bacterial cell

Q24:
- muscle contraction
- cell division
- mitosis
- protein synthesis
- transmission of nerve impulses
- active transport

Q25: cytoplasm

Q26: pyruvate

Q27: carbon dioxide and water

Q28: 2

Q29: control

Q30: temperature/size of paper disc/length of time the disc is soaked/spacing of the discs

Q31: 7

Q32: Any one from:
- *E. coli* does not produce amylase.
- *B. subtilis* produces the most amylase.
- *B. subtilis* produces more amylase than *H. hispanica*.
- *H. hispanica* produces less amylase than *B. subtilis*.
- *H. hispanica* and *B. subtilis* produce amylase

Q33: To make the results more reliable.

Answers to questions and activities for Unit 2

Topic 1: Producing new cells

The sequence of events of mitosis (page 83)

Q1:

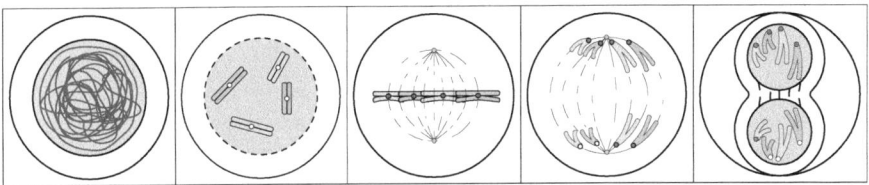

Cell structure (page 83)

Q2: Nucleus

Importance of mitosis: Questions (page 83)

Q3: To prevent the loss of genetic information and ensure the daughter cell carries out the exact same functions as the original parent cell.

Q4: Diploid

Q5:

Kangaroo cell type	Number of chromosomes
sperm	6
skin	12
nerve	12
zygote	12

Stem cells in animals: Questions (page 86)

Q6: Undifferentiated animal cells that can divide to produce more stem cells.

Q7: Embryonic stem cells AND adult stem cells.

Q8:

- **Embryonic**: Have the potential to divide and become almost any type of cell in the body.
- **Adult stem cells**: Have more limitations that embryonic stem cells. They are normally restricted to only being able to form the cell types from the tissue or organ they came from.

© HERIOT-WATT UNIVERSITY

Specialisation of cells: Questions (page 88)

Q9: b) tissue → organ → system

End of topic test: Producing new cells (page 89)

Q10: b) Specialised cells which have the ability to divide and produce new stem cells

Q11:

Pairs of chromatids are pulled apart	Chromosomes shorten and thicken
Chromosomes move to the equator of the cell	Chromosomes move to the equator of the cell
Cytoplasm divides	Pairs of chromatids are pulled apart
Chromosomes shorten and thicken	Nuclear membrane forms
Nuclear membrane forms	Cytoplasm divides

Q12: Spindle fibres

Q13:

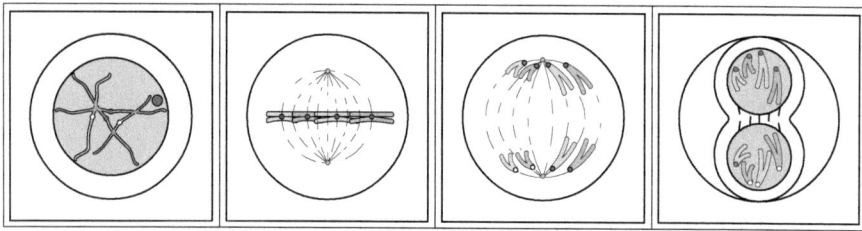

Q14: b) Fibres contract to pull chromatids apart to the opposite poles of the cell.

Q15: 7 times

Topic 2: Control and communication

Interactivity on parts of the brain (page 95)

Q1:

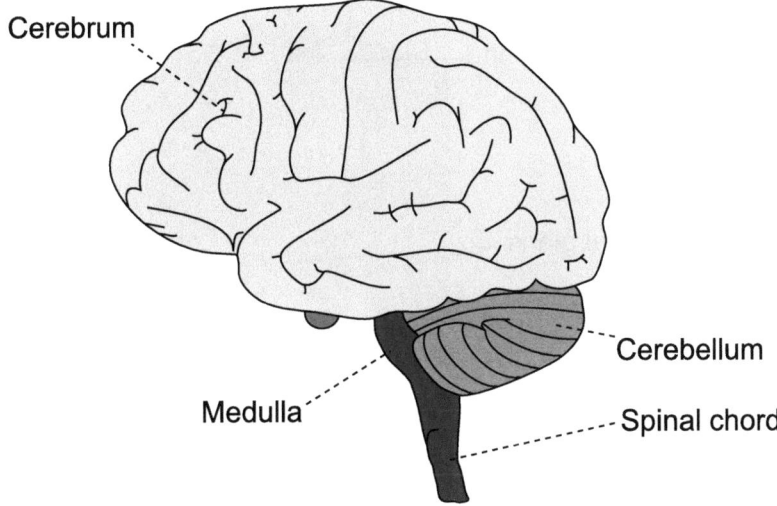

Nervous control: Questions (page 97)

Q2:

- Sensory neurons are nerve cells that carry electrical impulses from sense organs to CNS.
- Motor neurons are nerve cells that carry electrical impulses from the CNS to muscles and glands (effectors).

Q3: Brain AND Spinal chord

Q4: Sense organ → Sensory neuron → CNS → Motor neuron → Muscle

Q5:

1. The **cerebrum** is the section of the brain that controls memory, conscious thoughts, intelligence and emotions.
2. The **cerebellum** is the section of the brain that controls coordination, movements and balance.
3. The **medulla** is the section of the brain that controls breathing and heart rate.

Interactivity on the endocrine system (page 99)

Q6:

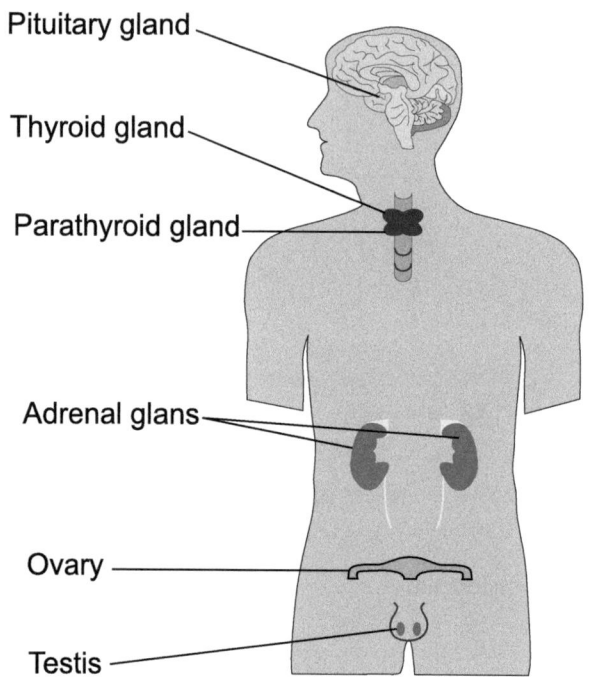

Hormonal Control: Questions (page 104)

Q7: Hormones are chemical messengers that transfer information from part of the body to another.

Q8: Endocrine glands

Q9: Pancreas

Q10:

Q11:

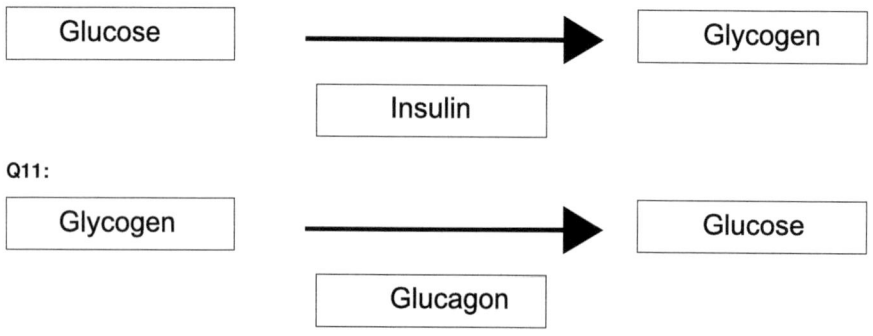

End of topic test: Control and communication (page 106)

Q12: Cerebellum

Q13:

- Receptors in the skin detect the stimuli.
- The electrical impulses are sent towards the spinal chord along sensory neurons to the inter neurones.
- Subject 'feels' the stimuli/ senses 'where' the needle is.

Q14:

- **Sensory neurons** are nerve cells that carry electrical impulses from sense organs to CNS.
- **Motor neurons** are nerve cells that carry electrical impulses from the CNS to muscles and glands (effectors).

Q15:

Stimulus → Receptor → Sensory neuron → Inter neuron → Motor neuron → Effector → Response

Q16:

- a) stimulus = pin in paw
- b) effector = muscle

Q17: The target cells in this tissue will have receptors that will recognises specific hormones and only bind to them

Q18: Endocrine gland

Q19:

- Option 1: Hormonal- chemical message; Nerve- electrical message
- Option 2: Hormonal- carried in blood/all over the body; Nerve- carried along specific pathways

Q20:

Receptors in the **pancreas** detect the increase in glucose levels and respond by producing the hormone **insulin**. This results in the conversion of **glucose** to **glycogen**.

Topic 3: Reproduction

Flower structure: Quiz (page 113)

Q1:

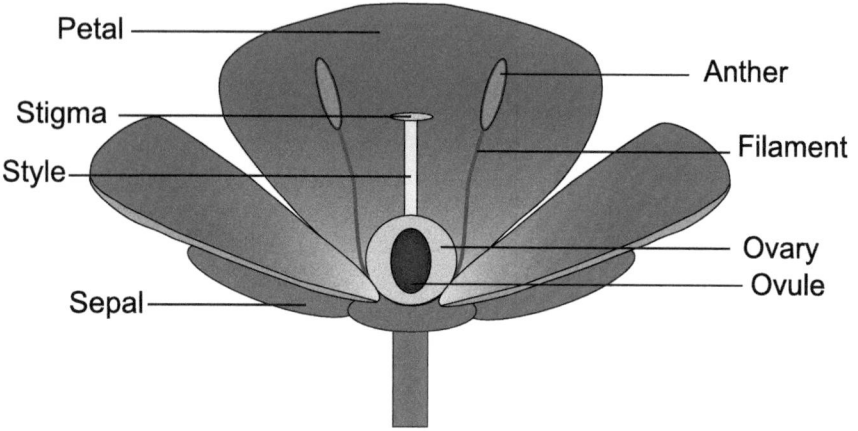

Flower structure: Questions (page 114)

Q2: Anther

Q3: Pollen grains

Q4: Ovary

Q5: Ovules

Reproduction in animals: Questions (page 116)

Q6: It has a tail that allows it to swim.

Q7: It has a large food store in its cytoplasm.

ANSWERS: UNIT 2 TOPIC 3

Fertilisation: Quiz (page 117)

Q8:

Term	Definition
Fertilisation	The fusion of two gametes
Gamete	Sex cell containing the haploid chromosome number
Zygote	A fertilised egg
Haploid	One set of chromosomes
Diploid	Two matching sets of chromosomes

End of topic test: Reproduction (page 119)

Q9: a) Pollen

Q10:

Term	Definition
Zygote	A fertilised egg
Diploid	Two matching sets of chromosomes
Haploid	One set of chromosome
Gamete	Sex cell containing the haploid chromosome number
Fertilisation	The fusion of two gametes

Q11:

Term	Definition
Ovary	Female organ in a flowering plant
Ovules	Structure containing the female gamete produced from the ovaries of plants
Pollen grains	Structure containing the male gamete produced from the anthers of flowers
Pollination	The transfer of pollen grains from an anther to a stigma
Anther	Organ in the flower that produces pollen grains

© HERIOT-WATT UNIVERSITY

Topic 4: Variation and inheritance

Variation in species: Questions (page 127)

Q1:

Term	Meaning
Discrete	Variations that are able to be categories into groups
Variation	Differences in the characteristics of members of the same species
Continuous	Variations that seen as one extreme to the other with a range of values in between

Q2: A group of interbreeding organisms who breed to produce fertile offspring.

Q3: The type of variation that shows characteristics ranging from one extreme to another.

Q4: The type of variation that shows characteristics that can be categorised into distinct groups.

Monohybrid crosses: Questions (page 132)

Q5: Thread-like structures containing DNA found in the nucleus of living cells.

Q6: 46

Q7: During gamete formation, **sex** cells receive different combinations of the **paired** chromosomes present in the original gamete mother cell.

Q8:

Term	Meaning
Polygenic	Type of inheritance involving several genes acting together
Phenotype	The physical appearance expressed by an organisms due to their genotype
Homozygous	Two alleles the same for a genotype i.e. AA or aa
Genotype	The particular alleles that an organisms has for a genotype
Recessive	The form of a gene which will only be expressed if the genotype is homozygous
Heterozygous	Two different alleles of a genotype i.e. Aa or Bb
Allele	The form of a gene

ANSWERS: UNIT 2 TOPIC 4

Interactivity: Punnet squares (page 132)

Q9:

Parental genotypes	h	h
H	Hh	Hh
H	Hh	Hh

Q10: a) discrete

Q11: b) heterozygous

Q12:

Parental genotypes	g	g
G	Gg	Gg
G	Gg	Gg

Interactivity: Phenotypic ratios (page 134)

Q13:

| 13 |

:

| 5 |

Q14: Two options:

- Fertilisation is a random process
- Numbers in sample too small

End of topic test: Variation and inheritance (page 136)

Q15: c) polygenic and show continuous variation

Q16: c) F_1 and F_2

Q17: D

Q18: Polygenic

Q19: A

© HERIOT-WATT UNIVERSITY

Topic 5: Transport systems of plants

Interactivity: Leaf structure (page 142)

Q1:

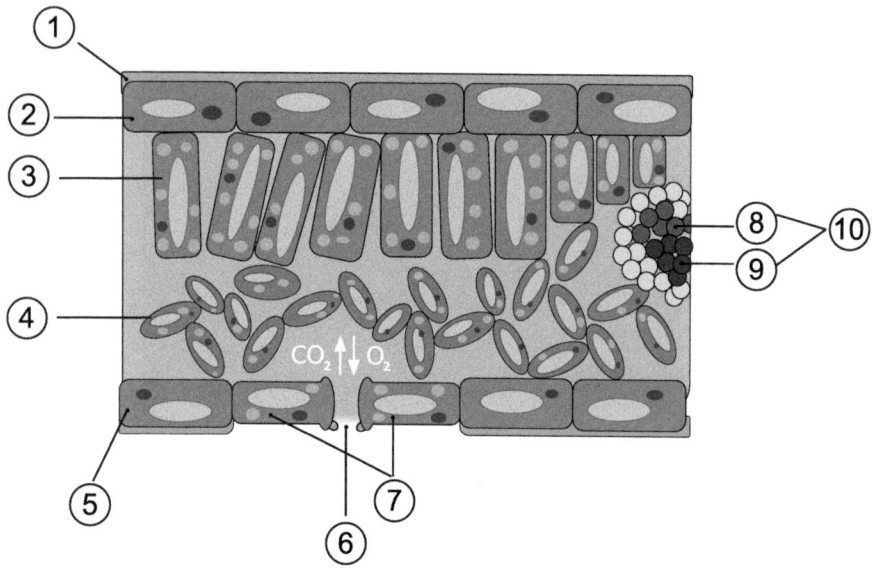

1		Waxy cuticle
2		Upper epidermis
3		Palisade mesophyll
4		Spongey mesophyll
5		Lower epidermis
6		Stomata
7		Guard cells
8		Xylem
9		Phloem
10		Vein

Interactivity: Estimating the age of a tree (page 147)

Q2: 11

Q3: 24

End of topic test: Transport systems of plants (page 150)

Q4: a) root hair > xylem > spongy mesophyll

Q5: Decrease

Q6: Increase

Q7: Increase

Q8:
- P has a greater number of stomata
- Q has fewer stomata

Q9: Guard cells

Q10: Water is **absorbed** by root hairs by **osmosis**. Water travels **upwards** in the **xylem**. Water travels to the **stomata** and **evaporates (or) transpires out**.

Topic 6: Transport systems of animals

Composition of blood (page 156)

Q1:

Blood component	Function
Plasma	Transports carbon dioxide, digested food, urea and hormones.
Red blood cells	Transports oxygen.
White blood cells	Ingests pathogens and produces antibodies.
Platelets	Involved in blood clotting.

Blood cells: Questions (page 157)

Q2: Haemoglobin

Q3: Image

Haemoglobin + Oxygen ⟶ Oxyhaemoglobin

Q4:

- a) They contain haemoglobin — a red protein that combines with oxygen
- b) They have no nucleus so they can contain more haemoglobin
- c) They are small and flexible so that they can fit through narrow blood vessels
- d) They have a biconcave shape to maximise their surface area for oxygen absorption

Immune system: Questions (page 159)

Q5: Disease causing micro-organisms

Q6: Phagocytosis

Q7: Each antibody is specific to a particular pathogen.

Pathways of blood through the body: Questions (page 162)

Q8:

Oxygen rich blood is said to be **oxygenated** and will then be pumped back to the heart so that it can be pumped out to the rest of the body. **Deoxygenated** blood is blood that lacks oxygen and is being returned to the heart from the body.

Arteries carry blood away from the heart towards an organ, while **veins** carry blood from an organ towards the heart.

ANSWERS: UNIT 2 TOPIC 6

Structure of the heart: Questions (page 165)

Q9:

Term	Meaning
Aorta	The main artery which carries blood away from the heart
Atria	Upper chambers of the heart that pass blood to the lower ventricles
Pulmonary artery	Artery that carries deoxygenated blood from the heart to the lungs
Pulmonary vein	Vein that carries oxygenated blood from the lungs to the heart
Vena cava	Blood vessels that carry deoxygenated blood to the heart from the body
Ventricles	Lower chambers of the heart that receive blood from the upper artria

Q10:

- right atrium
- right ventricle
- left atrium
- left ventricle

Q11: Aorta.

Q12: (Superior and/or inferior) Vena cava

Blood vessels: Questions (page 168)

Q13: Veins

Q14: Capillaries

Q15: Arteries

End of topic test: Transport systems of animals (page 169)

Q16:

Name of blood vessel	Diameter of central channel (mm)	Thickness of wall (mm)
Vein	25.0	1.25
Capillary	0.008	0.001
Artery	20.0	2.2

Q17: They have thinnest/thinner walls.

© HERIOT-WATT UNIVERSITY

Q18: b) X is the left ventricle which pumps blood to the body.

Q19: Coronary artery

Q20: Line B: right ventrice, pulmonary artery, pulmonary vein

Q21:

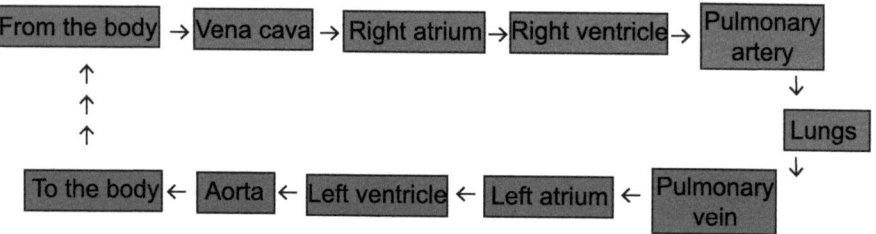

Topic 7: Absorption of materials

The need for transport: Questions (page 174)

Q1: Glucose OR amino acids

Q2: Urea

Q3: Carbon dioxide

Q4: Oxygen

The role of capillary networks: Questions (page 176)

Q5: Much more materials can pass by diffusion into cells.

Q6: Quick diffusion of materials into cells.

Q7: More opportunities for body cells to contact a capillary network.

The role of absorption surfaces in the body: Question (page 177)

Q8: b) high number of thin walled alveoli

Gas exchange in the lungs: Questions (page 180)

Q9:

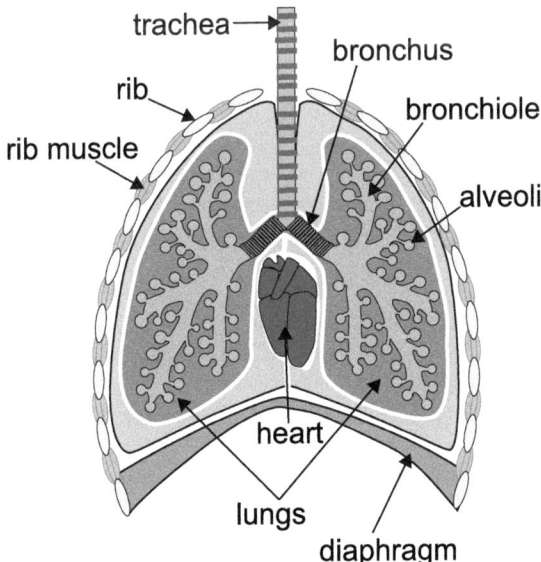

Absorption in the small intestine: Questions (page 183)

Q10: c) Amino acids and glucose

Q11: a) Fatty acids and glycerol

End of topic test: Absorption of materials (page 184)

Q12:

Cell type	Relative concentration of gases Oxygen	Relative concentration of gases Carbon dioxide
Red blood cell	Low	High
Alveolus cell	High	Low
Capillary wall cell	Medium	Medium

Q13: From cell of alveolus wall to cell of capillary wall to red blood cell

Q14: Any two from:
- Millions of alveoli for a large surface area
- Thin walled alveoli for quick diffusion of materials into cells
- Extensive blood supply around alveoli for transport

Q15:

Term	Meaning
Lacteal	Vessel in the villi that is responsible for transporting fats.
Lymph	Liquid that transports the products of fat digestion from the lacteal.
Peristalsis	Waves of muscular contractions that help food move through the alimentary canal.
Villi	Finger-like projections in the small intestine that provide a large surface area for absorbing food.

Q16:

Food travels through the alimentary canal by a process called **peristalsis**.

Muscles in the alimentary canal **contract** behind the food to squeeze it, while the muscles in front of the food **relax** to allow it to pass from one end of the canal to the other.

The food will come into contact with various enzymes in different parts of the digestive system until it reaches the **small** intestine.

Enzymes present throughout the digestive system aid the breakdown of large **insoluble** food molecules to small **soluble** molecules so that they can dissolve and diffuse across absorption surfaces.

Topic 8: Multicellular organisms test

Multicellular organisms test (page 188)

Q1:

Term	Meaning
Chromatid	Replicated copy of a chromosome.
Chromosome compliment	The number of chromosomes found in a cell.
Chromosome	Composed of DNA. Codes for all of an organisms characteristics.
Diploid	A cell that contains a double set of chromosomes.
Equator	Middle position of a cell where chromosomes align and attach to spindle fibres in mitosis.
Mitosis	A process of cell division that produces two genetically identical daughter cells.
Spindle fibres	Protein threads that pull chromatids apart during mitosis.

Q2:
Stem cells are **unspecialised** cells found in **animals**.

Q3:

- One use is to make more stem cells (self-renew);
- One use is to make cells that develop into specialised cells.

Q4: Function: To transport oxygen around the body.
Specialised structure(s): Do not have a nucleus and are biconcave in shape to increase the surface area

Q5:

Term	Meaning
Brain	Organ of the central nervous system that controls vital functions.
Central Nervous System (CNS)	Part of the nervous system made up of the brain and spinal chord.
Cerebellum	Section of the brain that controls coordination, movements and balance.
Cerebrum	Section of the brain that controls memory, conscious thoughts, intelligence and emotions.
Inter neuron	Nerve cell that are found in the CNS where they connect with other neurons.
Medulla	Section of the brain that controls breathing and heart rate.

© HERIOT-WATT UNIVERSITY

Q6:

Neurones	Function
Sensory	are nerve cells that carry electrical impulses from sense organs to CNS.
Inter	are nerve cells that are found in the CNS where they connect with other neurons.
Motor	are nerve cells that carry electrical impulses from the CNS to muscles and glands (effectors).

Q7:

- Anther — Where the male gametes are produced.
- Pollen grains — Male gametes.
- Ovary — Where the female gametes are.
- Ovules — Female gametes.

Q8:

STAGE	Description
Stage 1	The pollen grain begins to grow a pollen tube through the tissues of the style and towards the ovary.
Stage 2	The nucleus inside the pollen grain starts to make its way down the inside of the tube.
Stage 3	When the end of the pollen tube reaches and ovule in the ovary, it enters via a tiny hole.
Stage 4	The tip of the pollen tube will then burst to release the male gamete so that it may fuse with the female gamete and fertilisation can take place.

Q9:

Body cells are **diploid** in multicellular organisms and are **haploid** in gametes (sex cells).

Diploid means that each cell contains **two** sets of matching chromosomes.

Haploid means that each cell contains **one** set of chromosomes.

Q10:

Parental genotypes	r		r	
R	Rr		Rr	
R	Rr		Rr	

Q11:

- Continuous = Variations that seen as one extreme to the other with a range of values in between
- Discrete = Variations that are able to be categories into groups

ANSWERS: UNIT 2 TOPIC 8

Q12:

Term	Meaning
Genotype	The particular alleles that an organisms has for a genotype.
Heterozygous	Two different alleles of a genotype ie. Aa or Bb.
Homozygous	Two alleles the same for a genotype ie. AA or aa.
Phenotype	The physical appearance expressed by an organisms due to their genotype.
Polygenic	Type of inheritance involving several genes acting together.
Recessive	The form of a gene which will only be expressed if the genotype is homozygous.

Q13: Osmosis

Q14: The type of cells which control the opening and closing of stomata.

Q15:

Cell type	Function
Phloem	Sugar is transported up and down the plant in this living tissue.
Xylem	Water and minerals enter the plant through the root hairs and are transported in these.

Q16:

- (a) = xylem
- (b) = phloem

Q17:

- Z = left ventricle
- X = left atrium
- Z = right ventricle
- Y = right atrium

Q18:

Blood component	Function
Plasma	Transports carbon dioxide, digested food, urea and hormones.
Red blood cells	Transports oxygen.
White blood cells	Ingests pathogens and produces antibodies.

Q19:

Feature	Purpose
Large surface area	Much more materials can pass by diffusion into cells.
Thin walls	Very quick diffusion of materials into cells.
Extensive blood supply	More opportunities for body cells to contact a capillary network.

Q20:

Term	Meaning
Alveoli	Tiny sacs for gas exchange in lungs.
Cartilage	Flexible tissue in the trachea to keep the airway open.
Lacteal	Vessel in the villi that is responsible for transporting fats.
Lymph	Liquid that transports the products of fat digestion from the lacteal.
Villi	Finger-like projections in the small intestine that provide a large surface area for absorbing food.

… *ANSWERS: UNIT 3 TOPIC 1*

Answers to questions and activities for Unit 3

Topic 1: Ecosystems

End of topic test: Ecosystems (page 206)

Q1:

- predator
- consumer
- carnivore

Q2: producer

Q3: Any one from:

- oak tree → aphid → ladybird → blackbird
- oak tree → caterpillar → blackbird → hawk
- oak tree → caterpillar → thrush → hawk
- oak tree → woodbeetle → thrush → hawk
- oak tree → snail → thrush → hawk

Q4:

- aphids and caterpillar
- aphids and woodbeetle
- aphids and snail
- caterpillar and woodbeetle
- caterpillar and snail
- woodbeetle and snail
- blackbird and thrush

Q5: If the aphids were killed by a pesticide the number of ladybirds would be likely to **decrease** due to **lack of food OR loss of food OR less food**.

Q6: A group of organisms which can interbreed and produce fertile offspring.

Q7: Niche

Q8: interspecific

Q9: prey

Q10: An ecosystem consists of all the organisms (the **community**) living in a particular **habitat** and the non-living components with which the organisms interact.

Q11:

© HERIOT-WATT UNIVERSITY

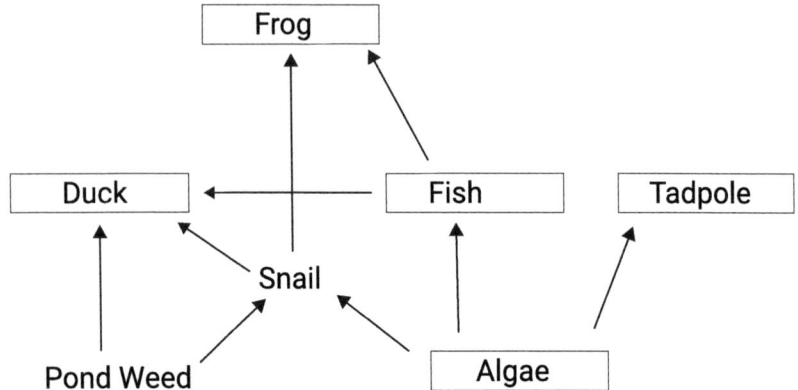

Q12: Ducks eat both plant and animal matter and are therefore described as **omnivores**.

Q13: increase

Explanation:

1. More food to eat OR more pond weed to eat OR increased food source OR increased food supply
2. Less predation by ducks OR less/fewer eaten by ducks

ANSWERS: UNIT 3 TOPIC 2

Topic 2: Distribution of organisms

Extended response (page 222)

Q1:

Light intensity:

1. Measured using a light meter.
2. The sensor is held upward and a reading is taken from the scale on the meter.
3. The user and/or other bystanders may inadvertently shade the light sensor.
4. Ensure bystanders stand well back to avoid shading the light meter.

Soil moisture:

1. Measured using a moisture meter.
2. The probe is pushed into the ground and a reading is taken from the scale on the meter.
3. There may be moisture left on the probe in between readings.
4. The probe should be wiped in between readings.

Soil pH:

1. Measured using a pH meter.
2. The probe is pushed into the ground and a reading is taken from the scale on the meter.
3. There may be soil left on the probe in between readings.
4. The probe should be wiped in between readings.

Temperature:

1. Measured using a thermometer.
2. The thermometer is placed in the experimental area, allowed to stabilise and a reading is taken from the scale.
3. When using a thermometer direct sunlight / heat from the user's hand can result in inaccurate readings.
4. Ensure the thermometer is set up in the shade / not held by the user.

End of topic test: Distribution of organisms (page 223)

Q2:

Abiotic factors	Biotic factors
Light intensity	Predation
Temperature	Competition
Soil pH	Disease

Q3:

Instrument: **light meter**

Description of use: **The sensor is held upward and a reading is taken from the scale on the meter.**

© HERIOT-WATT UNIVERSITY

Q4:

Instrument: **thermometer**

Description of use: **The thermometer is placed in the experimental area, allowed to stabilise and a reading is taken from the scale.**

Q5:

Instrument: **pH meter**

Description of use: **The probe is pushed into the ground and a reading is taken from the scale on the meter.**

Q6:

1. Wingspan greater than 35mm: go to **2**.
   Wingspan less than 35 mm: go to **3**.
2. **Eyespots OR Eyespots present**: go to 4
   No Eyespots OR Eyespots absent: go to Brimstone
3. Main wing colour blue: Common blue
   Main wing colour orange: Small heath
4. Main wing colour white: **Small white**
   Main wing colour orange: **Gatekeeper**

Q7: Quadrats are thrown randomly and the number of squares containing the organism of interest are counted.

Q8: Group 2 repeated the experiment OR group 2 used three quadrats rather than one OR Group 2 calculated averages for each species

Q9: Pitfall trap

Q10: Indicator species

Q11: water mite

Q12: flatworm

Topic 3: Photosynthesis
Extended response (page 232)

Q1:
1. Light energy trapped by chlorophyll
2. Water is broken down / split
3. Oxygen is released / diffuses (from the cell)
4. ATP / hydrogen is produced

Q2:
1. Controlled by enzymes
2. ATP provides energy / broken down / split
3. Hydrogen added to / joined to / used with carbon dioxide
4. Glucose / sugar is produced

End of topic test: Photosynthesis (page 235)

Q3: light reactions

Q4: 8

Q5: traps light energy

Q6: hydrogen or ATP

Q7: carbon fixation

Q8:

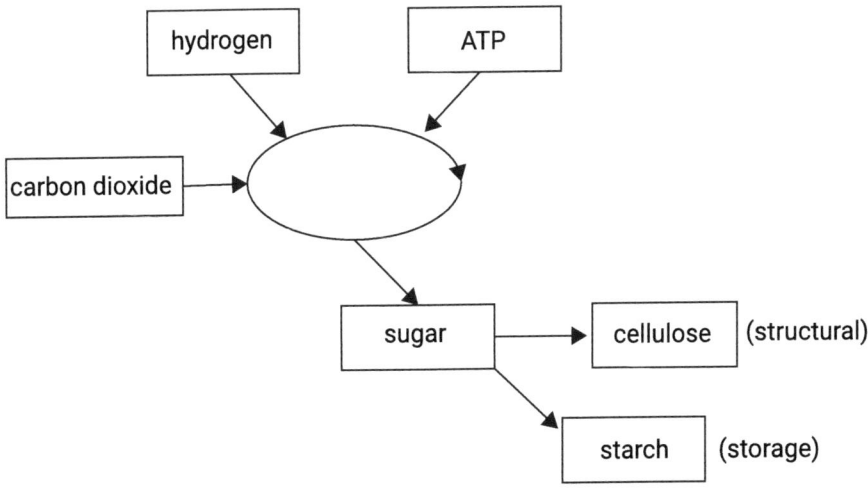

Q9:

Carbon fixation / photosynthesis is controlled by / requires / needs enzymes (1 mark)
(At high temperatures) enzymes are denatured / do not work (1 mark)

Q10: light intensity (both words required)

Q11: carbon dioxide concentration

Q12: temperature

Topic 4: Energy in ecosystems

End of topic test: Energy in ecosystems (page 243)

Q1: A diagram which shows the number of organisms at each level of a food chain.

Q2: pyramid of energy

Q3: growth

Q4: heat or movement or undigested material

Topic 5: Food production

Extended response (page 253)

Q1: Any 4 points for 4 marks:

1. This will increase algal populations/causes algal blooms.
2. Algal blooms reduce light levels, killing aquatic plants.
3. Dead plants/dead algae, become food for bacteria which increase greatly in number.
4. The bacteria use up large quantities of oxygen.
5. Reducing the oxygen availability for other organisms (causing them to die).

End of topic test: Food production (page 255)

Q2: Fertiliser

Q3: To make amino acids OR to make protein

Q4: Pesticide or insecticide

Q5: Bioaccumulation

Q6: biological control OR genetically modified crops OR GM crops

Q7:

1. Fertilisers leach into fresh water, adding extra, unwanted nitrates.
2. Extra nutrients cause algal populations to increase.
3. Overgrowth of algae reduce light levels, killing aquatic plants.
4. Dead plants, as well as dead algae, become food for bacteria which increase greatly in number.
5. The bacteria use up large quantities of oxygen, reducing the oxygen availability for other organisms.

Q8: algal bloom

Topic 6: Evolution of species

Answers from page 265.

Q1:

1. Part of a population becomes isolated/separated by a barrier.
2. Different mutations occur in each subpopulation/group.
3. Natural selection occurs OR selection pressures are different in each group OR advantageous mutations are selected for.
4. Subpopulations / groups are no longer able to interbreed to produce fertile offspring.

End of topic test: Evolution of species (page 267)

Q2: (Random) change to genetic material.

Q3: radiation OR X-rays OR UV light OR gamma rays OR chemicals OR colchicine OR mustard gas OR benzene

Q4: Allows population to adapt to changing (environmental) conditions OR makes it possible for population to evolve in response to changing conditions

Q5:

- In a large population of bacteria exposed to an antibiotic a random mutation occurs which confers resistance to the bacterium.
- Bacteria possessing the mutation survive as they have a selective advantage.
- Resistant bacteria pass on the favourable alleles (for antibiotic resistance) that confer the selective advantage.
- Antibiotic resistance alleles increase in frequency within the population.

Q6: natural selection OR survival of the fittest

Q7: Factors which can affect the survival or reproduction of an organism such as the presence of antibiotics in the example above are known as **selection pressures**.

Q8:

1. Different mutations occur in each sub-population.
2. Each sub-population evolves until they become so genetically different that they are two different species.

Q9:

Behavioural	Geographical	Ecological
a group of females selecting for a different male characteristic	river	pH
populations adopting different courting patterns	mountain range	salinity

© HERIOT-WATT UNIVERSITY

Topic 7: Life on Earth test

Life on Earth test (page 270)

Q1: Any from:
- giraffe and rhino;
- rhino and grasshopper;
- rhino and mouse;
- rhino and impala;
- grasshopper and mouse;
- grasshopper and impala;
- mouse and impala.

Q2: If the impala were killed due to over hunting the number of leopards would be likely to **decrease** due to **lack of food OR loss of food OR less food**.

Q3: A group of organisms which can interbreed and produce fertile offspring.

Q4: Biodiversity

Q5: Interspecific

Q6: Predator OR consumer OR carnivore

Q7: Indicator species

Q8:

1. Antennae present .... go to **2**
   Antennae absent .... go to **3**
2. **Three pairs of legs**..................... go to 4
   Six pairs of legs........................ Water louse
3. Length greater than 100mm .... Sludgeworm
   Length less than 100mm .... Rat-tailed maggot
4. Length greater than 15mm .... **Mayfly nymph**
   Length less than 15mm .... **Stonefly nymph**

Q9: pH is **an abiotic** factor.

Q10:

pH meter
Put the probe into the water and read the pH from the scale on the meter.

Q11: Pitfall trap

Q12:

Name: **Light reactions**
Location: **Chloroplast**

Q13:
A: **Oxygen**
B: **Hydrogen**

Q14: ATP

Q15: Carbon fixation

Q16: Hydrogen and ATP are used with carbon dioxide to produce sugar.

Q17: Light intensity

Q18: Temperature

Q19: A diagram which shows the number of organisms at each level of a food chain.

Q20:
Irregular shapes of pyramids of numbers based on different body sizes can be represented as true pyramids of **energy**.

Q21: Heat or movement or undigested material.

Q22: To make amino acids OR to make protein

Q23: Genetically modified crops OR manure

Q24:

1. Stage 1: Fertilisers can leach into fresh water, adding extra, unwanted nitrates
2. Stage 2: **Algal population increase OR algae bloom.**
3. Stage 3: Algal blooms reduce light levels, killing aquatic plants.
4. Stage 4: **Bacteria population increase OR bacteria increase in number.**
5. Stage 5: The bacteria use up large quantities of oxygen, reducing the oxygen availability for other organisms.

Q25: Bioaccumulation

Q26: Biological control OR genetically modified crops

Q27: (Random) change to genetic material.

Q28: Natural selection OR survival of the fittest.

Q29: The light form were the best adapted individuals so they survive to reproduce, passing on the favourable light coloured alleles that confer the selective advantage. These alleles increase in frequency within the population.

Q30: One of:

- Geographical
- Ecological

© HERIOT-WATT UNIVERSITY

- Behavioural

Q31: To make the results more reliable.

Q32: Light intensity

Q33: Size/diameter/mass/weight/thickness/surface area of leaf OR type of leaf OR concentration of carbon dioxide solution OR volume of carbon dioxide solution OR temperature OR distance of lamp.

Q34: 28%

Q35:

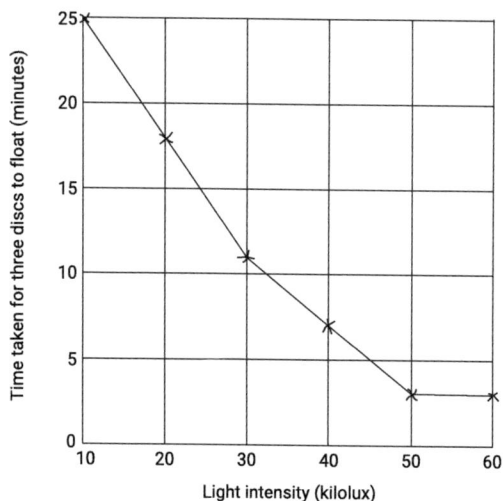

Q36:

As the light intensity increases, time taken for three discs to float decreases up to 50 kilolux.

After that, time taken for three discs to float remains steady/ stays the same.

Q37: 5

Q38: To find out if chemical fertiliser affects plant growth/height differently compared to organic fertiliser.

Q39: Plant growth/height

Q40: Control

Q41:

Treatment	Plant height (cm)
No (additional) fertiliser	20
Chemical fertiliser	27
Organic fertiliser	24

Q42: Chemical fertilisers improve plant growth/increase plant height more than organic fertiliser.

Q43: Only used lettuce plants/too small a sample.

BV - #0028 - 240122 - C0 - 210/148/19 - PB - 9781911057857 - Matt Lamination